The HAZOP Leader's Handbook

The HAZOP Leader's Handbook

How to Plan and Conduct Successful HAZOP Studies

PHIL EAMES

ELSEVIER

Elsevier
Radarweg 29, PO Box 211, 1000 AE Amsterdam, Netherlands
The Boulevard, Langford Lane, Kidlington, Oxford OX5 1GB, United Kingdom
50 Hampshire Street, 5th Floor, Cambridge, MA 02139, United States

Notices
Knowledge and best practice in this field are constantly changing. As new research and experience broaden our understanding, changes in research methods, professional practices, or medical treatment may become necessary.

Practitioners and researchers must always rely on their own experience and knowledge in evaluating and using any information, methods, compounds, or experiments described herein. In using such information or methods they should be mindful of their own safety and the safety of others, including parties for whom they have a professional responsibility.

To the fullest extent of the law, neither the Publisher nor the authors, contributors, or editors, assume any liability for any injury and/or damage to persons or property as a matter of products liability, negligence or otherwise, or from any use or operation of any methods, products, instructions, or ideas contained in the material herein.

ISBN: 978-0-323-91726-1

For Information on all Elsevier publications
visit our website at https://www.elsevier.com/books-and-journals

Publisher: Susan Dennis
Acquisitions Editor: Anita Koch
Editorial Project Manager: Moises Carlo P. Catain
Production Project Manager: Niranjan Bhaskaran
Cover Designer: Greg Harris

Typeset by MPS Limited, Chennai, India

Dedication

To Amanda

For her patience and support throughout the writing of this book.

Contents

Preface

In 2016 I was fortunate to be commissioned by Matt Stalker at the Institution of Chemical Engineers to redesign and subsequently deliver the Institution's *HAZOP Leadership & Management* training course, which was re-launched successfully in 2017.

Over many years as a Hazard and Operability (HAZOP) leader, I had experienced a wide range of practices in respect of how the methodology was applied, matched by a wide variation in leadership styles. But more recently, since 2010, I had had the privilege of working within a number of organisations that were making great efforts to apply the methodology to a high standard and insisting on carefully selecting their HAZOP leaders, not just according to their experience but for their ability as facilitators and their preparedness to apply the highest standards.

In my research prior to redesigning the course, I concluded that the practices I had experienced and deployed within these organisations were far more rigorous and advanced than what I found in the literature, and this was confirmed by discussions with other like-minded HAZOP leaders. That was the genesis of this book. I wanted to impart my experiences to the widest possible audience and bring the literature up-to-date with current best practices, as far as I understand it.

During the redesign process, in October 2016, the US Chemical Safety Board (CSB) released its final report (https://www.csb.gov/williams-olefins-plant-explosion-and-fire-/) into a 2013 explosion at an olefins plant in Geismar, Louisiana in which 2 men died and 167 others were injured. Among a number of process safety management failures was the failure of three PHA (HAZOP) studies over a period of 12 years to address the hazard of an off-line reboiler being heated while isolated from its thermal relief valve (blocked in), which included at least one study in which the hazard was identified but the HAZOP recommendation to address it was not written in a way that would have helped to increase the likelihood of it being acted upon.

Reading the report on this incident galvanised me in my efforts to highlight the importance of applying the HAZOP methodology thoroughly and the responsibility of the HAZOP leader for making sure that happens. That is what this book is about.

Acknowledgements

I would like to thank Matt Stalker at the Institution of Chemical Engineers for the opportunity to redesign and deliver the Institution's HAZOP Leadership & Management training course, which was the genesis of this book, and Tracey Donaldson, also at the Institution of Chemical Engineers, for encouraging me to approach Elsevier to publish it.

CHAPTER 1

Introduction

1.1 Reputation and reality

Since its introduction in the UK chemical industry in the late 1960s the use of Hazard & Operability (HAZOP) Study has expanded across all sectors of the process industries and all parts of the world, so that today it is the pre-eminent hazard identification tool used in process hazards analysis. Very few reputable organisations would undertake the development of a new process facility today without undertaking a HAZOP study during the detailed design phase of the project; many organisations have applied HAZOP retrospectively to their processes and, indeed, revalidate or repeat studies on a regular basis throughout the life of their facilities. The technique has established an impressive reputation as an effective methodology for the identification of process hazards. But what is the reality? How effective is it? The answer is that we don't really know, because the effectiveness of a HAZOP — the extent to which it identifies and describes all the hazards — is difficult, if not impossible, to measure. But what we do know is that there is evidence to suggest that we may have become complacent about how effective it is. Fig. 1.1, in which the overarching term process hazards analysis includes HAZOP, is based on an analysis of the causes of the 100 largest ever process industry loss events [1].

It shows that in more than half of the largest loss events in the short history of the process industries there were weaknesses in the management systems designed to identify process hazards. Having seen such weaknesses at close hand throughout my career in operations and process safety, I would suggest that there is no room for complacency.

Because HAZOP is a facilitated group exercise to identify process hazards in a creative way, it is self-evident that the effectiveness of the application of the methodology is heavily dependent on the facilitator of the study: you, the HAZOP leader.

This book is designed specifically to help HAZOP leaders plan and execute successful HAZOP studies, based on many years of experience of

The HAZOP Leader's Handbook
DOI: https://doi.org/10.1016/B978-0-323-91726-1.00010-3

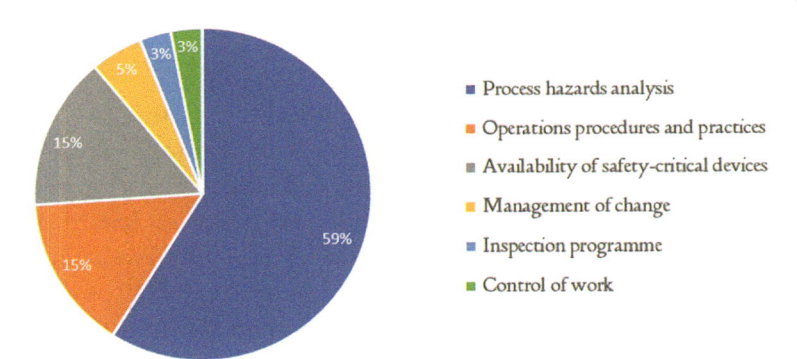

100 Largest Losses
"Non-Mechanical Integrity" Failure Losses
Secondary Management System Failures

- Process hazards analysis
- Operations procedures and practices
- Availability of safety-critical devices
- Management of change
- Inspection programme
- Control of work

Figure 1.1 Process hazards analysis failure featured in 59% of major loss events [1].

participating in, observing and facilitating HAZOP studies, as well as observing, training, mentoring and assessing HAZOP leaders.

The book assumes that the reader has an understanding of the methodology and has experienced HAZOP meetings; it does not seek to explain the methodology itself, which is straightforward and explained in the (relatively small) HAZOP literature [2−4]. Rather, it focuses on the application of the methodology and the responsibilities and skills of the HAZOP leader in the preparation, execution and reporting of the study. Although the principal subject is HAZOP, much of the content equally applies to other facilitated hazard identification techniques such as Hazard Identification (HAZID) or Failure Modes Effects and Criticality Analysis (FMECA).

1.2 Why this book?

Imagine it's the morning of the first meeting of a new HAZOP that you are to lead. You have arrived on the site that belongs to an organisation with which you are not familiar and, although you have had several discussions with the sponsor of the study, you have not been able to meet any of your team members in advance. You are nervous as usual, but excited. You know you have prepared as well as you could and you are looking forward to the challenge of leading and motivating a new team and guiding them through, what you hope, will be a thorough and effective study. You are happy with the meeting room and you have checked out the projector. This has calmed the nerves a bit, but you know how important the team is to the success of the study, so you are hoping that they'll arrive enthusiastic and well-prepared. You want to make a good impression right from the start.

Ten minutes before the agreed starting time for the meeting, the first team member arrives. Your relief is palpable; maybe we'll make an efficient, on-time start! As you approach the newly arrived team member you watch her disinterestedly sit down and take out her phone. You approach and, with a short cough to get her attention, introduce yourself as energetically as possible as the HAZOP leader. She smiles politely and greets you with her name, but then her attention starts to move back towards her phone. To try and prevent this you start a conversation.

'Have you done much HAZOP work before'?

'Yeah, we have to do them for all the projects we do, so I've had to come to meetings from time to time'.

'How have you found the experience'?

'Pretty boring really, but I suppose it's necessary. How long is this one meant to take? I've got some important work going on right now an another project'.

'I've estimated that we should be able to do it in about 10 days of meetings'.

'Oh, really? (slight sigh of disappointment). . .10 days, as much as that'?

'Yes, we need to make sure we do as good a job as possible'.

'You sound like you enjoy it. You do a lot of this then'?

'Yes, I spend about half of my time on HAZOP facilitation. I do enjoy it'.

'Wow, I don't know how you can do it. . .'

Here we go again. . .it looks like this study is going to be another challenge in terms of motivating the team to achieve an effective outcome. It's a challenge that HAZOP leaders face on a regular basis.

Over many years' experience in a wide variety of organisations I know that HAZOP leaders have experienced this type of reaction on many occasions. It concerns me because it suggests a lot of people consider HAZOP to be boring: a necessary evil that we have to undergo from time to time to meet the requirements of the process safety management system or get to the next stage of the project. Worse, it could be indicative of lack of value for the technique. It is exemplified by the saying widely attributed to the late process safety guru Trevor Kletz [5]: 'No one jumps out of bed on a Monday morning shouting "Hooray, I've got a HAZOP today!"'. Given that HAZOP is intended to be an inclusive and creative process to understand — in depth — how the process works and, of course, how we can lose control leading to hazardous or unproductive situations, this sort of view undermines the reputation of HAZOP and the threatens the effectiveness of the methodology; overcoming it presents an enormous challenge to HAZOP leaders, who do jump out of bed regularly on a Monday morning, perhaps not shouting but at least thinking, 'HAZOP today, I'm up for it!'

So the prime purpose of this book is to provide guidance specific to HAZOP leaders to help you to maximise the effectiveness of your HAZOP studies, get the most benefit from the methodology, promote consistency and rigour in its application, sustain its well-earned reputation and, hopefully, enhance your own, of course.

There are three other reasons why I think this book is necessary: first, to preserve the integrity of the methodology by resisting the increasing pressure to cut down on study time; second, to promote the application of current best practice — which in the view of many practitioners and organisations has moved ahead of that described in the HAZOP literature [2–4]; third, to increase the emphasis of facilitation skills as a core competence for HAZOP leaders. Let me explain each of these in turn.

First, the integrity of the methodology. When I first experienced HAZOP in the 1980s, in the pre–computer era of course, meetings were recorded entirely by hand, usually in pencil on A3 sheets of paper. Not only was this slow and laborious, but team members were not able to see how the discussions were being recorded. So meetings were mostly held on mornings only, to enable the afternoon to be spent by the leader and recorder editing and typing up the record sheets. These would then be copied and reviewed by the team at the beginning of the next morning's session. Not very efficient, but then we didn't have an alternative. And it never seemed to be a problem; there always seemed to be enough time to progress through the study at a measured pace without delaying the project schedules or disrupting team members, the project manager or site management. Over the years we've overcome the issue of recording on paper, and it's hard to imagine a HAZOP study being done today without it being recorded using software and projected 'live' to the team. However, during this time other pressures have been mounting, chiefly the relentless drive for lower costs and improved efficiency across the process industries. So today, in my experience, it is very rare that a HAZOP is undertaken without time and resource constraints, which are an ever-present threat to the rigour and effectiveness of the study. The focus of much effort in a lot of organisations has been on executing studies faster and more efficiently, and terms such as 'Coarse HAZOP', and 'Preliminary HAZOP' [6] are now regularly used to describe what are, in a lot of cases, superficial studies carried out on incomplete designs. The emphasis seems to me to be more on efficiency than thoroughness or effectiveness; in my view the prevalence of this type of exercise threatens to undermine the power of the methodology and devalues the term HAZOP; if the term HAZOP has to be used to describe them, then perhaps they should just be called 'bad HAZOP'. This book aims to preserve the integrity of the traditional methodology by unashamedly focusing on its effectiveness.

Second, the promotion of best practice. In recent years, as part of the drive to improve process safety management, a number of organisations have significantly increased their focus on HAZOP, in particular the rigour with which the methodology is applied. Examples of this are in requirements for more comprehensive process safety information to support the study, evidence of the competence of the HAZOP leader and team, the style and detailed recording of scenarios and the structuring of recommendations for further risk reduction. In my view, this has taken the application of the methodology beyond that described in the current literature (at least the most commonly used texts [2−4]). This book aims to share some of these developments in practice and − hopefully − prompt the HAZOP community to develop and share further improvements in future.

And finally, a HAZOP leader is much more than just the chair of a meeting; the HAZOP leader is a project manager and a facilitator. Although some organisations bring in HAZOP leaders with the sole task of chairing the meetings (not a role I have enjoyed, personally), in my experience this is not the best way to achieve an effective study. Far better that the HAZOP leader is involved from the conception of the study (the identification of the need for it) right through to the delivery of the HAZOP report and even − although this is much less prevalent these days with the frequent use of contracted HAZOP leaders − to the completion of the recommendations arising from the study. In this way, the role of the HAZOP leader is much like a project manager, steering the study from conception through a preparation phase, an execution phase (the HAZOP meetings) and then a reporting phase. And even within the execution phase, the HAZOP leader is much more than a meeting 'chair', because this term implies that the leader is simply guiding the meeting participants through an agenda. Of course, HAZOP isn't like that. It's about the application of a structured methodology, but in an imaginative and creative way, and so the HAZOP leader in this role is a facilitator; someone who not only understands how to apply the methodology but can inspire and motivate a team to work creatively to achieve an effective study that is not tedious or boring. No mean feat!

This book will take you on a journey through the HAZOP process. It starts with the understanding of the enormous challenge that we are taking on as HAZOP leaders in Chapter 2. Then, if that doesn't put you off, it moves on to the planning of the study in Chapter 3. Execution of the study is covered in two parts: the technical application of the methodology in Chapter 4 followed by the role of the HAZOP leader as a facilitator in Chapter 5. In Chapter 6 we discuss the effectiveness of the technique and how we might be able to assess it and improve it. Chapter 7 completes the process in the form of the development of the final report. The book concludes with a summary of ways in which you can improve your performance as a HAZOP leader.

References

[1] R. Jarvis, A. Goddard, An Analysis of Common Causes of Major Losses in the Onshore Oil, Gas and Petrochemical Industries, Loss Prevention Bulletin 255, Institution of Chemical Engineers, 2017.
[2] F. Crawley, B. Tyler, HAZOP Guide to Best Practice, third ed., Elsevier, 2015.
[3] F. Crawley, B. Tyler, Hazard Identification Methods, European Process Safety Centre, 2003.
[4] T. Kletz, HAZOP and HAZAN, fourth ed., Institution of Chemical Engineers, Rugby, 1999.
[5] P. Coutts, The Chemical Engineer Issue 920, Institution of Chemical Engineers, 2018, p. 58.
[6] R. Wittkower, B. Singh, A. Botto, M. Hull, P. Jukes, 21st International Offshore and Polar Engineering Conference, Maui, Hawaii, USA, 2011.

CHAPTER 2

More than just a chair!

After more than 50 years of Hazard & Operability (HAZOP) practice some organisations are still using the term 'HAZOP chair'. I suppose we can be thankful that they've stopped using this outmoded term in its gendered format, but when I see a request or advertisement for a HAZOP chair the alarm bells start ringing because there is much, much more to be being a HAZOP leader than chairing the meetings. Indeed, the role of the HAZOP leader in the meetings — as we'll explore in this chapter — is much more than just chairing them!

So the use of the term 'chair' may indicate that the organisation does not fully appreciate the role of the HAZOP leader in the HAZOP meetings. However, it may be even worse than that: it may indicate that the organisation does not fully appreciate the important role of the HAZOP leader throughout the process of conceiving, planning, executing and reporting the study. This involvement of the HAZOP leader throughout the lifecycle of the study is a major theme of this book and one of the critical success factors for achieving a successful outcome from the study. So if you see an advertisement for a HAZOP chair, don't jump in without first having a serious discussion of your project manager's or client's expectations.

Before we examine the process and the role of the HAZOP leader in more detail, let's take a look at the responsibilities of a HAZOP leader. Because understanding these responsibilities leads directly to the importance of the leader being involved at every stage in the life of the study.

2.1 The HAZOP leader's responsibilities

As a HAZOP leader, what are you responsible for? The most obvious answer is to get the study done: to deliver the study on time and in full to make sure that the full scope of the study is completed in the agreed time. We could add 'at the agreed cost' in relation to using no more than the agreed level of resources. Getting behind schedule and having to hold more meetings than originally planned in order to meet the deadline carries some form of cost: financial if it involves additional contractor time,

7

for example, or opportunity cost if it diverts staff away from their normal responsibilities. The experience of knocking on the project manager's or client's door and having to beg for more of his or her peoples' time is not often a good one. That's the first hint that it's useful for the HAZOP leader to be involved in the planning process: even if you yourself didn't estimate the amount of time required for the study, it still reflects badly on you when you have to ask for more resource or time.

So, delivery is our primary objective. But is that all? What about the quality of the study? We'll talk a lot more in Chapter 6 about how we can (or cannot) measure the effectiveness or quality of a HAZOP study, but as a HAZOP leader you should be concerned about the depth and thoroughness of your team's work: the extent to which you identified, described and analysed the process hazards. Not only that you covered the full scope, but that you covered all parts of it to an appropriate degree of thoroughness. When you submit the final report, you want to have the confidence to say that the team did the best job it could have done. You don't want to have nagging doubts that you might have skimped parts of the study — the utilities or product storage areas for example — because you were running out of time.

When we think about delivery and quality — thoroughness or effectiveness — together it quickly becomes obvious that they are intimately related: reduce the thoroughness, increase the speed. Balancing pace and thoroughness is a constant concern for the HAZOP leader. If the pace feels slow, are we going into too much detail? If the pace feels fast, are we challenging the design strongly enough? And of course you're performing this mental balancing act at the same time as you are facilitating... and thinking about what's coming next...and thinking about how the team are performing and feeling. So it's often a source of stress, which might lead to the temptation to sacrifice thoroughness for speed as the deadline approaches!

We'll come back to the team in a moment. But first, when we talk about quality or effectiveness we also have to think about the application of the methodology. There are two aspects to this. The first is compliance with the requirements of the organisation's specific HAZOP methodology if it has one, as many larger organisations do. As HAZOP leaders we need to make sure that we meet the organisation's standards, which might include the composition of the team, the way nodes are constructed, the set of deviations that are applied and the way the worksheets are filled out (e.g. the way equipment or safeguards are described). It may cover more technical aspects such as the extent to which non-return valves or pressure relief devices are acceptable as safeguards, or how the organisation's risk matrix is deployed to assess scenarios. For contracted HAZOP leaders in client organisations this can pose a headache when you are facilitating a study in one company one week and another the next. Once again it points towards the importance of being involved in the preparation phase, during which you will need time to understand the specific requirements of the methodology. Often the client will have an audit checklist against which it will examine the report, and this will give you valuable information in relation to the client's expectations.

The second aspect of the application of the methodology is the wider professional one. Regardless of an organisation's specific requirements, as professional engineers we have a duty to apply the HAZOP technique thoroughly, not only to provide the organisation with a high-quality study but to uphold the reputation of the HAZOP technique itself. We could grandly call this upholding our professional (and even personal!) integrity. This issue can often arise when you go into an organisation to lead a HAZOP for the first time. You might review a previous HAZOP report for the facility in question and find it superficial, either in terms of the way it has been documented or the thoroughness with which the study was conducted, or both! Or, in the early meetings you might find members of the team happy to accept the validity of a pressure relief valve as a safeguard without checking that the device has adequate capacity for the scenario in question, or happy to document the presence of a non-return valve that doesn't appear on the piping and instrumentation diagram (P&ID). Or, in one experience I had when I agreed to attend site for the study believing that preparations had been made, happy to embark on a HAZOP study without any P&IDs! This is yet another pointer towards having involvement from as early a stage as possible in the preparation, and preferably even the conception, of the study.

And finally, HAZOP leader, your team. Yes, it's your team and nobody else's. So what is the nature of this responsibility? Obviously, you want to get the best possible performance out of the team as the primary means of achieving a successful study; that's the main goal of your role as a facilitator. But is there more? I've already mentioned that we often hear that HAZOP has a reputation for being boring; not for us as leaders of course because we are highly motivated in the art of facilitation and spend every minute of the study completely absorbed in it, but for the members of the team. The experience of each of the individual team members is

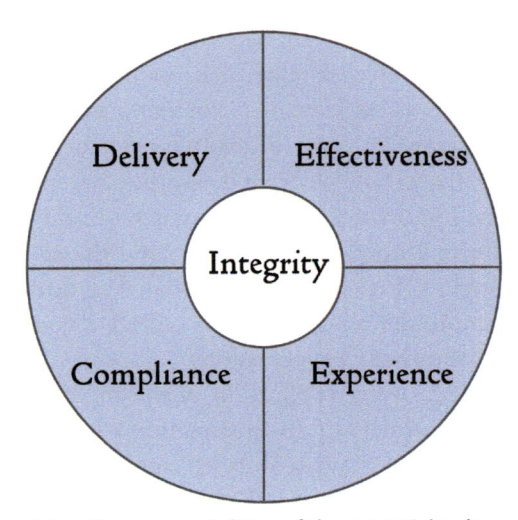

Figure 2.1 Summarising the responsibilities of the HAZOP leader.

the thing that will determine their enthusiasm for participating in future studies and their view of HAZOP as useful or ineffective: a necessary evil or a valuable use of time; fascinating or boring. So I think we have a responsibility to our team members to make their experience as positive as possible. This will not only lead to better studies but will maintain — and hopefully enhance — the reputation of the technique. And again, the earlier we are involved in the planning process, the more opportunity we might have to get to know the team members beforehand (even help to select them) and to start securing their commitment, earning their trust and building the team.

Fig. 2.1 summarises the responsibilities of the HAZOP leader discussed above.

While facilitating the study you'll be constantly thinking of these five responsibilities and how the study is going in relation to each of them. This is one reason why you'll end up exhausted at the end of most HAZOP days!

2.2 Are you ready for the challenges?

Now that we're clear about the awesome responsibilities you have as a HAZOP leader, let's discuss some of the challenges that you'll face. Box 2.1 presents the definition of the HAZOP technique, taken from the international standard for HAZOP IEC61882 [1]:

Before we introduce some reality into this definition, one omission is worth mentioning; there is no mention of facilitation. HAZOP is carried

---

Box 2.1 A definition of HAZOP.

A HAZOP study is a detailed hazard and operability problem identification process, carried out by a team. HAZOP deals with the identification of potential deviations from the design intent, examination of their possible causes and assessment of their consequences.

From [1].

Box 2.2 A more realistic definition of HAZOP.

A HAZOP study is a detailed hazard and operability problem identification process, carried out **within difficult time and resource constraints, on a design that may not be fixed and with incomplete information,** by a team, **members of which may have never met one another and have different degrees of ownership and interest in the process, led by a facilitator who may have little facilitation experience, and for a customer whose expectations may not be clear.**

HAZOP deals with the identification of potential deviations from the design intent, examination of their possible causes and assessment of their consequences, **in a creative brainstorming process, the effectiveness of which depends on the manner in which the technique is applied, but which cannot be measured in any scientific way.**

---

out by a team facilitated by an experienced leader. Without this, the definition reads like a straightforward technical methodology, when in fact the effectiveness of the methodology is determined by the effectiveness of the team. You might say it could be implied in the above definition, but I think the absence of the word 'facilitate' is certainly worth highlighting.

Now for some reality from my own experiences to illustrate some of the challenges we will face. In Box 2.2 I have adjusted the IEC definition by adding the words in bold text, including emphasis on facilitation, and possibly just a little exaggeration.

Let's discuss in turn each of the six challenges I've added to the definition.

2.2.1 Time and resource constraints

We all appreciate that in today's dynamic business world almost everything we do is subject to time and resource constraints and HAZOP, as a time-consuming and resource-intensive activity, often attracts attention. In the context of a design project the HAZOP study is almost always on the project's critical path, which guarantees that there will be pressure to complete it on time, and often — if the project timings have slipped — to compress it. This in turn puts the HAZOP leader under pressure and turns the spotlight on the delivery/quality relationship unless scope can be brought into the equation to avoid an impact on quality caused by compressing the time allowed (much more of this in Chapter 6). In the context of ongoing operations (HAZOP revalidation or re-do) time may be less of a constraint in itself, but the costs of the study will be borne by the operations organisation, which may in turn lead to pressures to compress the timescale.

From a resource standpoint the most important aspect is normally the availability of key people, in both project and ongoing operational contexts. Any depletion of the team, even for short periods, represents a threat to the quality of the study and this is something of which the HAZOP leader must remain acutely aware.

2.2.2 Design completeness

In an ideal world the HAZOP study for a capital project should be conducted when the design has been finalised, typically when the P&IDs are, at or almost at, the 'approved for design' stage. However, as we have said, HAZOP is invariably on the project's critical path and there is almost always a reluctance to pause the detailed design process to await the outcome of the HAZOP. So HAZOP is often done on an incomplete design; the important factor is the extent of its incompleteness. Large numbers of 'bubbles', hold points or queries on the P&IDs, or blank boxes around vendor packages or other sub-units are the tell-tale signs. If the design is not sufficiently complete the study will quickly encounter difficulties due to uncertainty and this can lead to long debates within the meetings and large numbers of recommendations being generated to deal with the fall-out. In the worst case it may become apparent that the design is just not very good, with obvious omissions such as isolation valves, instrumentation or destinations for vent or drain points. In either case, the HAZOP meeting can quickly turn into a design review. If such shortcomings are not uncovered before the HAZOP meetings start, then as HAZOP leader you will be in the difficult position of having to consider stopping the meetings and postponing the study. Another difficult conversation with the project manager.

2.2.3 Quality of information

P&IDs are obviously the focal point for HAZOP, but a thorough study also requires a large amount of supporting information to facilitate discussions, for example Cause & Effect Diagrams to understand how protective systems operate, equipment datasheets to confirm design pressures and pressure relief philosophies to understand whether relief devices are appropriate for the hazardous scenario under discussion. Absence of important information like this can severely slow down the meetings and will frustrate the team; it quickly leads to the generation of large numbers of recommendations aimed at confirming the information required to correctly document the hazardous scenarios. Alternatively, it could lead to incorrect assumptions being made, for example that a relief device has been sized for a gas blow–by scenario, which could lead to a scenario being misrepresented and appearing better protected than it really is.

2.2.4 Facilitation and group dynamics

In the history of HAZOP the overriding, sometimes the only, job requirement for the role of HAZOP leader has been engineering experience. But if we are really serious about getting the maximum value out of the HAZOP technique which, after all, consumes an enormous amount of time and resource, then we need to make sure that HAZOP leaders are good facilitators. Anyone who has been in a HAZOP meeting will have experienced some of the challenges of facilitation: dealing with team members who don't really want to be there; getting people who have never met one another before to work together; handling team -members with 'pet' interests or interests driven by their role or discipline; dealing with people who are naturally shy, dominating or aggressive. There are many, many more. Those readers who have facilitated a HAZOP meeting will already know this and you will have your own war stories (and hopefully some success stories too). The importance of facilitation skills is a major feature of this book.

2.2.5 Customer expectations

As HAZOP leaders we have a very clear understanding of the technique, how we like to apply it and what the final product — our report — should look like. It may be tempting to assume that our customer — the organisation's project manager or the client's process safety engineer — has the same understanding, but this would be a dangerous assumption. As I said when discussing the integrity of the HAZOP methodology in Chapter 1, different practices have developed in relation to the application of HAZOP and there are wide variations between organisations in terms of what is understood as HAZOP, from rigorous to relatively superficial. So we need to take care to understand exactly what our customer wants when they ask for a HAZOP study and be prepared to 'educate' them if their expectations are anything less than what we would consider as recognised good practice or if they are seeking to use HAZOP for purposes to which it is not suited. 'Can you come and lead a 1-day HAZOP?' always makes me suspicious!

2.2.6 Measurement of effectiveness

There is a widespread assumption that a thorough HAZOP study should identify the vast majority of process hazards within the facility or design being studied. However, can that be measured? Can we state with any certainty that we have undertaken an effective or successful HAZOP? In reality we can't: we can try to identify all the significant hazards but we can never be sure that we haven't missed some...the 'unknown known' scenarios that we haven't been rigorous enough to uncover but are likely to accompany the technology, as well as the 'unknown unknown' scenarios that we have not been sufficiently creative to imagine. As HAZOP leaders we are constantly asking ourselves, 'What might we have missed?'

Having reviewed the responsibilities of being a HAZOP leader and the challenges you will face, we now need to complete the picture by discussing the subject of competence: what combination of qualifications, skills and experience do you need to have to fulfil the responsibilities, face the challenges and lead successful studies?

2.3 Qualifications, experience and skills

The role of a HAZOP leader is a relatively complex one, combining technical knowledge, engineering and operational experience and, last but by no means least, facilitation skills.

Let's start with qualifications in so far as they relate to the technical knowledge required to lead HAZOP studies. Some large companies have defined competence criteria which specify that a HAZOP leader should hold a first degree in an engineering discipline – typically chemical or process engineering – or even a second process safety qualification, such as an MSc. or Professional Process Safety Engineer status. This is fine but may exclude candidates that could perform as excellent HAZOP leaders. The important factor here is the capability of the HAZOP leader to confidently facilitate discussions relating to process plant design and operation, which may require getting involved in some technical aspects of these discussions, asking questions or prompting further explanations. I have worked with very competent HAZOP leaders that were degree-qualified in pure sciences or, in some cases, not degree-qualified at all but had a technical background; the keyword is technical. However, of greater importance is experience.

HAZOP leaders should have a significant amount of 'hands on' experience in the operation of process facilities, preferably hazardous ones; many HAZOP practitioners would probably consider 10 years as a minimum. It is this experience that will equip a HAZOP leader with the understanding of how processes can go out of control, how unexpected problems can arise, the myriad ways in which equipment and control systems can fail to operate as designed, and the way in which safeguards may not operate on demand, procedures are not kept up-to-date or not followed, and humans fail to do what we expect them to do (or what they themselves expected to do!). It is this experience that enables the HAZOP leader to encourage the team to think of what could go wrong, to ask questions that help to develop potential hazardous scenarios to their ultimate consequences, and to challenge assumptions that processes and equipment will always operate as expected. We could say such experience develops a healthy scepticism that enables the HAZOP leader to encourage creativity and avoid complacency in the team (with an armoury of 'war stories' from themselves and others to stimulate and sometimes amuse the team when the situation is appropriate).

So, a technical background and operational experience are fundamental requirements. But to what extent should this training and experience relate directly to the technology or process industry sector that is the subject of the HAZOP study? It has been accepted wisdom for some time in most process industry sectors apart from nuclear, that the HAZOP leader does not have to have experience in the specific technology being studied. and this has often formed part of the HAZOP leader's opening introduction to the study ('I don't have a background in this industry sector but I am a competent HAZOP leader with experience across a range of sectors', usually said in the most earnest way possible). However, it is becoming more common, particularly in the oil and gas sector, for organisations to expect a HAZOP leader to have some knowledge of the technology or process, to give them confidence that the HAZOP leader can make sure that the appropriate questions are raised in the study. It is certainly helpful to go into an oil and gas HAZOP with an understanding of terms like gas blow-by and hydrate formation or to a water treatment HAZOP with an understanding of chlorine re-liquefaction; not to do so would probably damage your credibility from the outset. I have certainly on occasion felt exposed leading a HAZOP in a technology or sector not previously familiar to me but have found that some homework and famil-iarisation with the process before the study, for example by reviewing a previous or related HAZOP report, are often enough to enable you to establish an appropriate level of credibility with the team. It is important that you appreciate the main hazards associated with the technology under study.

A second aspect of the experience requirements is obviously that of experience of using the HAZOP technique itself. Industry bodies have thus far stopped short of developing a qualification standard for HAZOP leaders because the coaching and experience components are so important (and always lie outside of their capability to influence). However, a num-ber of organisations offer well-respected training courses for HAZOP leaders, normally of 1 week duration, and this should be considered as pre-requisite for developing as a HAZOP leader. This should preferably be preceded by attendance at a number of studies and followed by atten-dance at more studies as a recorder and then as a mentored leader to build up an experience base before 'flying solo'.

And last, as stated earlier, but not least...facilitation skills. Throughout the history of HAZOP, leaders have tended to come from an engineering background. But why should an engineer necessarily be a good facilitator of teams? Recall the TV sketch [2] in which Dilbert is taken to the doctor by his mother, concerned at his obsessive aptitude for taking apart and fixing things. He is diagnosed as having 'The Knack, a condition charac-terised by an extreme intuition about all things mechanical and electri-cal...and utter social ineptitude'. When his mother asks whether he can lead a normal life, the doctor replies, 'No, he'll be an engineer'. Facilitation skills have rarely been part of HAZOP training courses, other than a chance to practice as a leader in short classroom exercises, but these are the skills that the HAZOP leader needs to be able to get the most out of the team and therefore maximise the effectiveness of the study. They include listening, supporting, summarising, challenging and conflict resolution and require an understanding of group dynamics, team development and individual communication styles. I believe it is so important that Chapter 5 is devoted to this subject.

2.4 Developing and demonstrating your competence

I have mentioned that many larger operating organisations and some process safety consulting organisations have defined competence criteria for HAZOP leaders; these often form part of a training, competence development and assessment programme that enables the organisation to maintain a pool of competent HAZOP leaders that are 'accredited' (qualified and approved). The best of these programmes involve a period of apprenticeship in which the aspiring leader is mentored and coached by a respected leader as they develop their experience and facilitation skills before being formally assessed, normally by a panel of experienced HAZOP leaders or process safety specialists. Assessment can take a number of forms, including observation of HAZOP sessions and inspection of final reports from studies which the candidate has led. Box 2.3 summarises some selection criteria:

Depth of experience includes both technical and operational aspects, while width of experience refers to the practice of HAZOP in the context of other process hazard identification and risk assessment tools. Professional standards refer to the ability to facilitate decision-making in the absence of full technical understanding.

An example of a detailed protocol for the assessment of HAZOP leader competence is presented in Appendix 1.

If you are fortunate enough to be able to develop as a HAZOP leader within one of these programmes, great. Otherwise, it will be up to you to

---

BOX 2.3 Some important aspects of HAZOP leader competence.

- Qualifications
- Depth of experience
- Width of experience
- Knowledge of the HAZOP methodology and recognised good practice
- Professional standards
- Knowledge of industry and/or company engineering codes and standards
- Understanding of the relevant regulatory environment
- Personal competence and facilitation skills

plan your own development. If you aspire to leading HAZOP studies as a third party in other organisations, expect to be subjected to an assessment like the one presented in Appendix 1. In any event, it is useful as a HAZOP leader to be able to describe your competence, whether it be selling yourself in an interview situation for a job as a contracted leader or establishing your credibility with a new team as part of the introductory presentation you make at the start of a study. To help with this, consider developing a Training and Experience Log, which details your qualifications, the training you have received as a HAZOP leader and the experience you have of HAZOP studies as a team member, recorder and leader. It should capture — as a minimum — a list of the studies detailing your role, the technology or type of facility, the type of study (continuous, batch or a mixture), the dates within which it took place and the duration. You may also want to include experience of other process hazard identification studies you have experienced, such as Hazard Identification (HAZID) and Failure Modes Effects and Criticality Analysis (FMECA). It is always useful to try and retain copies of the reports from the studies you have led, as some organisations may ask for samples of HAZOP worksheets. Finally, make sure you keep any positive comments you have received from team members or study sponsors that you can use as references.

2.5 Make sure you are independent!

Although it is not part of your competence, while we are discussing your suitability as a HAZOP leader it is worth mentioning the importance of an independent facilitator as a critical requirement for any HAZOP study. Using a third-party (contracted) HAZOP leader will almost always guarantee the required independence unless the person in question has previously worked for the organisation or if your consultancy is involved in other projects within the organisation, which might complicate things a little. Deploying internal leaders requires more care. In general, recognised good practice would suggest the more independent the better, so if you are requested to lead a HAZOP on a different site or in a different division of your company, or on a project with which you have no involvement, then that is likely to be fine. But independence becomes more difficult to justify if you are requested to lead a HAZOP on the site or within the project on which you are working. With the best will in the world, in these circumstances it may be difficult to provide a suitable degree of challenge within your team because you are part of the same organisational culture. On a personal level, if you know the members of the HAZOP team, or have worked with them before, then that could make it more difficult for you to challenge them (perhaps subconsciously). On a technical level, if your organisation considers single, untested non-return valves as adequate protection against reverse flow, then it will be difficult for you to challenge this within the HAZOP. From the perspective of risk, if your organisation has a reluctance to rank hazardous scenarios as 'high risk' on account of a subconscious 'group think' effect that fears attracting negative attention from senior management, then you are unlikely to be able to challenge this type of thinking effectively. As an independent HAZOP leader working with a site-based or project-based team you are likely to encounter situations like this where you think to yourself, 'I can't believe they think this is safe enough'! Your task is then to persuade them to challenge their own assumptions, perhaps using your own experiences or those from incidents that you are familiar with.

2.6 Still want to be a HAZOP leader?

Big responsibilities, significant challenges, lots of experience and practice required. Who would want to be a HAZOP leader? Fortunately, many of us still do and have developed practices, tools and behaviours that enable us to maximise the likelihood that our studies will be effective and successful. The aim of the rest of this book is to follow the HAZOP process from the conception of the study through to the delivery of the final report, describing current good practice, together with tools and techniques that have been developed over many years of experience of managing HAZOP studies. The aim is to challenge you in relation to your experiences of HAZOP up to now and to help you to develop and improve your leadership skills in future.

References

[1] IEC 61882:2016, Hazard and Operability Studies (HAZOP Studies) — Application Guide, British Standards Institute, 2016.
[2] Available at https://www.aiche.org/chenected/2010/04/dilberts-mother-realizes-her-son-destined-become-engineer-video.

CHAPTER 3

Fail to prepare, prepare to fail

In Chapter 2 we saw that there is much more to be being a Hazard & Operability (HAZOP) leader than simply chairing the meetings, hence my dislike of the term 'HAZOP chair', which implies that someone — albeit experienced in applying the technique — can just turn up and start chairing a HAZOP study. Although doubtless this has been tried, the best HAZOP leader in the world would struggle to go into a study 'cold'. Anybody who has had to join a study as a short-notice substitute for another leader knows that it is incredibly difficult and detrimental to the quality of the study, at least for the time it takes to get up to speed in terms of familiarity with the process, understanding the way the study had been facilitated before you joined it, and understanding and adapting to the group dynamics. Needless to say this time is not a good experience for team members either.

We need to appreciate the role of the HAZOP leader throughout the process of conceiving, planning, executing (that's the meetings part) and reporting a HAZOP study. This is the process of undertaking a HAZOP that is described in the international standard for HAZOP Application IEC 61882:2016 [1] and is summarised in Fig. 3.1.

The more involvement the HAZOP leader has in the definition and preparation of the study, the more effective the study is likely be. This involvement enables the leader to influence and support the aims of the study and then help to make sure that sufficient time and resources are put in place to maximise its effectiveness. The less you are involved in the preparation phase, then the more likely you will be a hostage to fortune in terms of whether the study will be a success.

That's why we are devoting a whole section of this book to the conception and preparation phase. The contents of the section are supplemented by the Preparation Guide presented as Appendix 2, which provides important considerations to take into account across five phases of a study: preliminary, scoping, terms of reference, execution and reporting.

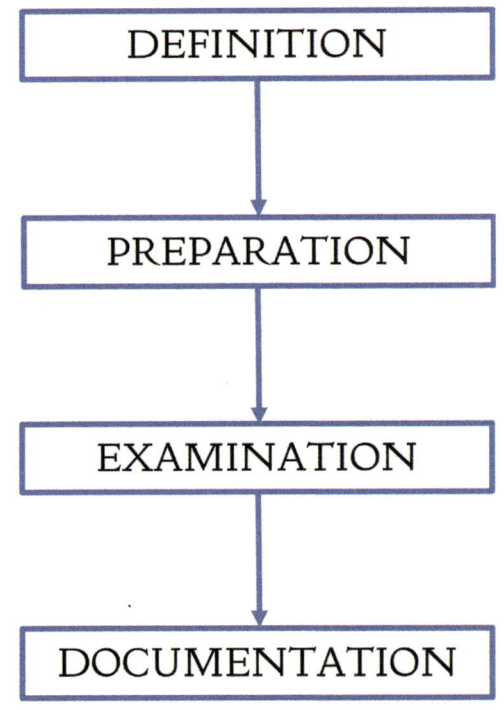

Figure 3.1 The HAZOP process. *From International Standard IEC 61882:2016, Hazard and Operability Studies (HAZOP Studies) — Application Guide, British Standards Institute, 2016.*

3.1 Is it HAZOP that you need?

Why do we start in a preliminary phase (or definition stage, as the international standard terms it) before a decision to conduct a study has been made? Surely it's not the HAZOP leader who decides to conduct the study? (much as many a consultant would like!). The reason for starting your involvement at the conception of the project is to enable you to use your experience to help the client or sponsor decide if HAZOP is the right tool to satisfy their needs and then, if it is, to help frame the specific scope and objectives of the study and, thereafter, to work to make sure that their expectations are met. Of course, it may not be possible for you to be involved at such an early stage, especially if you are contracted as a third party, but ideally you should be involved. If you are brought in during preparations or, worse, simply to facilitate the meetings, then you had better make sure that you are comfortable that you will be able to meet the client or sponsor's expectations before your take on the job!

If you are fortunate enough to be involved or consulted at the conception stage then this is your opportunity to understand what the client or sponsor really wants to achieve, remembering from Chapter 1 that you should not assume that this person has a good understanding of what HAZOP is all about. In the case of a major project, a significant modification or the re-HAZOP of an existing facility (all of which are likely to be required by the organisation's process safety management system then it is likely that the organisation does indeed want a detailed line-by-line examination of the process to identify hazards associated with deviations from design intent: a proper HAZOP (but beware requests for 'coarse' or 'preliminary' studies — see below). In this case, it will be relatively straightforward to proceed, provided that you are happy that the design is at a sufficient stage of completeness, that the scope is clear, and that you are happy that the organisation understands the time, resource and financial implications.

In other cases HAZOP, with all the time and resources it takes to do it properly, may not actually be the right tool to achieve what the client or sponsor wants. An example might be a design that is actually component-based equipment rather than a flowing or step-by-step process, such as a compressor system, hydraulic control unit or chemical dosing pump skid. In such cases a technique like Failure Modes Effects and Criticality Analysis (FMECA) [2] may be more suitable, or even Structured What If? Analysis (SWIFT, sometimes called What If?/ Checklist, a typical checklist for which is provided in Appendix 3). The important thing is to understand the nature of process and ask yourself, is HAZOP the best way forward here?

Just as important as the type of process is the reason for selecting HAZOP. Over the years I and others have been asked to lead HAZOP studies on processes for which HAZOP is appropriate, but for reasons such as the plant is experiencing operational problems, reliability problems or has had a number of safety incidents. Are these good reasons for going to the expense of conducting HAZOP? A more appropriate response would probably be to analyse the problems themselves, individually and collectively, in order to identify the underlying issues; there is no guarantee that HAZOP would provide the right answers to these issues. In discussions with the client or sponsor over the objectives of the study it is important to remember the limitations of HAZOP as a technique: it assumes that the process is well designed; it analyses single, isolated deviations and assumes that the rest of the process is not in a degraded state; it assumes operational and maintenance competence; it considers risk in a qualitative and judgemental way; it is unlikely to identify causal patterns typical of process safety incidents.

Assuming that the client is sure that they want HAZOP, and for the right reasons, do they mean a proper, line-by-line detailed HAZOP or do they want a 'coarse' or 'preliminary' study as a precursor to doing a full study at a later date? (Or not doing a full study!) This has become more prevalent in recent years in some major projects as, knowing it is often on the critical path, organisations seek to reduce the overall HAZOP time by holding 'preliminary' or 'coarse' studies before the design is complete on the assumption that the full HAZOP, when it is eventually done, will be quicker as a result of the up-front work. While I can understand the desire to avoid project delays caused by the length of HAZOP studies, I have seen no evidence that this approach has been successful because, at the end of the day, a line-by-line full HAZOP still needs to be performed in the normal way and the availability of information from earlier 'preliminary' or 'coarse' studies that relate to an unfinished design may not be fully relevant; moreover, reviewing this information requires additional time in itself.

The risk in undertaking 'coarse' or 'preliminary' studies is that you end up leading a superficial study that should not really be called HAZOP. Do you really want to be involved in that? If an organisation wants to undertake hazard identification as early as possible, then one obvious way of doing this is to use the Hazard Identification or HAZID technique [3], which can be deployed as soon as the Process Flow Diagram or PFD is available. Another approach would be to deploy a form of Structured What If? (SWIFT) study. This is arguably likely to produce a similar output to a 'preliminary' or 'coarse' HAZOP without the reputation of HAZOP being undermined by its association with a superficial form of study. Appendix 3 shows a typical example of a SWIFT checklist.

One final point relating to discussions of the client's or sponsor's concerns or desires. If there are concerns raised that the design may not be sufficiently robust in some aspects or that the existing plant is experiencing problems or incidents, then this is valuable information for a HAZOP leader. You can use such information to make sure that the study covers the areas of concern and write the final report in a way that addresses the concerns directly. One way to do this is to present — in the summary of the study's recommendations — an analysis that draws attention to recommendations that relate to these concerns. If the quality of operating procedures is a concern raised before a re-HAZOP, then you can highlight a set of recommendations that relate to operating procedures. (This is discussed further in Chapter 7.) If you can do this then you are more likely to please the client or sponsor.

3.2 Can you plan for success?

Once you have established that a full HAZOP will satisfy the requirements of the client or sponsor, the next challenge is to make sure that the client understands what will be involved in terms of time, resource and cost. This is the starting point for managing the client's expectations and an important decision point for you in terms of whether you feel you will be able to meet these expectations while delivering a study that is conducted according to recognised good practice (and your own professional standards). Remember that the client may not actually understand what is involved in executing a good quality study. It is often the case that the organisation has already decided how long the HAZOP should take, and this may not have been on the basis of a proper estimate made by an experienced leader. 'Can you come and lead a HAZOP for us, it should only take a few days?' or 'We've set aside a week for the HAZOP' are common requests; it is important that you understand the basis for the anticipated duration and make your own estimate based on as much information as you can get: ideally a set of piping and instrumentation diagrams (P&IDs) that represent the scope of the study.

In relation to the cost of a HAZOP study, this obviously includes the cost of preparing for, performing and reporting the study, principally determined by the anticipated duration. However, it is also worth discussing with the client their expectations in terms of managing the recommendations from the study. It is likely that the cost of implementing recommendations will far exceed the cost of actually doing the study, but I have been surprised many times by clients that have such confidence in their designs that they believe HAZOP won't produce any significant recommendations or clients that haven't thought about how they might manage the output from the study.

Understanding the process by which the recommendations from the study will be managed may enable you to provide valuable assistance by making sure that recommendations are produced in a format that is helpful, for example using the organisation's standard HAZOP recommendation format or a format that allows the recommendations to be easily uploaded into the organisation's or project's action tracking database, if they have such a system.

Once you are comfortable that you understand the client's expectations, then you have to decide if you are able to meet them. Will it be possible to complete the study in the specified time to a standard that meets recognised good practice? If you are concerned about this then now is the time to raise it. You may be able to persuade the client to provide more time but, if you cannot do this, you have the decision to walk away or risk compromising your professional standards (and further damaging the reputation of the technique). If you are confident that you have sufficient time, then the next question is whether you can enter into a robust contract with the client in terms of a mutual agreement as to how the study is going to be conducted. This is often referred to as the terms of reference for the study, frequently abbreviated to ToR.

3.3 Terms of reference

Terms of reference (often called a charter in the United States) is an agreement between the client or sponsor of the study and the HAZOP leader that describes how the study will be conducted. In an ideal world it should be drawn up by the client or sponsor, but in reality it often falls to the HAZOP leader to do it. If your client shows no inclination to compose terms of reference, I would definitely recommend that you draft it and present it to them for discussion and agreement. This document is absolutely crucial: it should provide complete clarity as to how the study is to be executed and, very importantly, should provide you with a point of reference in the event that the study does not go according to plan and you are forced into difficult discussions with your customer. It is both a contract and an insurance policy should things not go as well as expected. I would emphasise that it is worth drawing up terms of reference for every study, regardless of size. Timescales can be more demanding for shorter studies where, for example, an extra day on a planned 4-day study represents 25% extra time, whereas longer studies inherently provide more catch-up time in the event that the programme slips. And, of course, the smaller the study, the simpler the terms of reference will be to prepare.

> **BOX 3.1 Key sections for terms of reference.**
> - Title and signatures
> - Background
> - Objectives
> - Scope
> - Methodology
> - Process safety information
> - Personnel
> - Schedule and deliverables
> - Report content and distribution

So what should be in terms of reference? A full list of recommended contents is provided in Appendix 4 and this can be used to construct a template for your document. Of course, organisations may have their own standard template, in which case Appendix 4 can be used as a checklist in the discussions you will have before terms of reference are drawn up, if you are lucky enough to be given the opportunity. Let's take a look at the main sections of the contents provided in Appendix 4 — summarised in Box 3.1 — and point out some important considerations.

3.3.1 Put the client's agreement front and centre!

It's always useful to have the agreement you have come to with the sponsor of the study solemnised in signatures and displayed prominently on the front cover of the document. Of course, the agreement can be displayed at the end of the document, but why not put it up front for further emphasis? Signatures are important as a signal of intent, which recalls a difficult experience. On the first day of a major HAZOP, as a third-party leader with a new team (some of whom had flown a long way to be there), we started with the introductions and preliminary presentation (discussed further in Section 3.9) during which I introduced the terms of reference, which had been agreed previously by an e-mail exchange with the client but not 'wet-signed'. One of the client's representatives was quick to challenge the absence of wet signatures and promptly refused to allow the study to start until he had a signed copy of the document in his hand. Unable to persuade him to allow us to continue and get the signatures later (I was only a contractor, remember, and he represented the client), we completely lost the best part of the first day and I lost a load of credibility with my new colleagues. . .a good example of how HAZOP meetings can blow up unexpectedly and why HAZOP leaders need to be constantly 'on their toes'. Interestingly, 4 days into the study, I had to ask the client to remove this individual from the team; not as revenge (much as that thought had occurred to me on that first morning) but because this individual turned out to be unable to discuss any aspect of process risk without demanding that recommendations be made to seek more and more technical information on which hazardous scenarios could be constructed. In spite of all my efforts, this completely paralysed the team to the extent we were making almost no progress; they were visibly demoralised by their fellow team member and colleague. I have not encountered such an individual since but it goes to show that a HAZOP leader must be ready to deal with an enormous spectrum of personalities and behaviours (much more of this in Chapter 5). From that study onwards, however, I have always insisted on signatures.

3.3.2 Describe the background or context of the study

It is useful to discuss the context of the study when trying to understand and clarify the purpose and objectives (and a good way to make sure that the team has this same understanding) and to summarise it at the front of the terms of reference. This may include a broad description of the facility to be studied in the context of the process operations and business as a whole (e.g. an offshore production platform in the context of field operations as a whole) or a description of the project of which the facility forms part (e.g. a catalytic cracker as part of a refinery expansion or a reaction unit as part of a pharmaceutical product development). Aspects such as the age of the facility, the scale of the facility and its operating history (for re-HAZOP studies) and the current status of the project (for new builds and modifications) are also important for you and your team to understand. For example, on an ageing facility you may want to direct specific attention to aspects such as equipment condition (corrosion for example) during the HAZOP sessions by virtue of your choice of deviations or the relative emphasis you give to them. If previous HAZOP studies have been undertaken (in the case of re-HAZOP) then some information on these studies and how they were followed up in terms of addressing recommendations is also useful. In the case of a new design, it is useful to describe the status of previous hazard identification studies (such as HAZID) conducted in the earlier stages of the project.

3.3.3 Be clear about your objectives

Although it can seem obvious ('do a HAZOP'), spelling out the objectives in more detail is a good test of how clear they are and why you are doing the study. In a project context, you are likely to be aiming to fulfil the requirements of corporate project management standards; for a re-HAZOP you may also be aiming to meet the requirements of corporate standards but there may be other factors or concerns such as the robustness of the process design or the performance or age of the facility. In either case, your objective is to subject the process to a detailed line-by-line analysis of deviations from design intent, with the aim of identifying improvements to the design or operation, thereby reducing risk and improving operability. There is never any harm in spelling out the purpose of HAZOP! It may also be appropriate to highlight that the study will aim to identify hazards that may have safety, environmental, asset damage or business (commercial) implications if only to emphasise that risk will be considered broadly. This brings to mind numerous occasions where I have been asked to consider the safety and environmental risk only, to the exclusion of operability. Operability is a core component of HAZOP (there's a clue in the name): to ignore it would be to undertake only a partial study as well as miss opportunities to improve the operation of the process, not forgetting that poor operability is often an underlying cause of process safety incidents by virtue of its leading to deviations from procedures, 'workarounds' and unauthorised modifications.

3.3.4 Carefully define the scope of the study

You cannot be too careful when describing the scope of the study. If this is not absolutely clear, then you will absolutely compromise your ability to deliver what the client or sponsor expects, and may well end up in dispute over the completeness of the study or the length of time it takes. You cannot make a good estimate of how long you will need unless you are completely sure of the scope.

A good place to start is with a description of the process or facility that you are required to study. But this is only a start because it will not provide sufficient detail. You will need to look outside of the process to define the boundaries of the study (e.g. battery limits, asset ownership boundaries or interfaces with upstream or downstream processes); you will also need to go inside the process or facility to list the main systems and units that will be included (the PFD is often useful here). There are two reasons for this. First, some of the systems or units that form the process may not need to be included: the client may not wish to study shared utility systems, raw material unloading or storage systems, product storage systems or vendor-supplied packages or may intend to study them separately at another time. It is very important to list not only the systems and facilities that will form part of the study but also to list specific systems that are to be excluded: clarity is everything.

When you have a clear list of included and excluded systems, the next challenge is to define the boundaries of these systems or the interfaces with other processes outside of the one you are to study; where are the specific boundaries or interfaces located on the relevant P&IDs and how can these be described? Again, the PFD can be helpful here but remember that these will be the start or end points for nodes and so it is vital that they are pin-pointed on the P&IDs as well. Establishing boundaries based on main process isolation points or battery limits is always helpful but not always possible.

The final part of scope description is to identify and list all the P&IDs that cover the process between the boundaries that you have defined. It is useful to include the list of P&IDs, along with their revision numbers, as an appendix to the terms of reference. This provides a clear definition of the study scope such that any additional P&IDs that may be identified later, or any more recent revisions that may emerge during the study, can be treated as scope changes, and the implications of studying them discussed and agreed with the client.

3.3.5 Specify how you will apply the methodology

The scope of the study is obviously crucial in relation to your agreement with the client or sponsor and, of course, your estimate of the amount of time that the study will take. The way in which you will apply the HAZOP methodology can also have an important bearing on the time required and so it is important to provide some detail in relation to this so that it is clear to the client or sponsor.

Start with the type of HAZOP you are to apply. Will it be continuous, batch or procedural, or a mixture? Employing batch or procedural HAZOP is likely to be more time-consuming than traditional continuous HAZOP on a node-for-node basis, so factor this into your time estimate (we'll discuss this in Section 3.5). Then make sure you agree on what modes of operation will be covered in the study. This cannot be overstated: how are you going to address start-up and shutdown for example? It may be appropriate to study these as stepwise procedures in addition to undertaking continuous HAZOP on the process, in which case you will need significantly more time. Many HAZOPs nowadays employ 'start-up' and 'shutdown' deviations within each continuous node, but this is relatively ineffective in comparison with studying the start-up or shutdown separately as a procedure (are 'start-up' and 'shutdown' really helpful deviations?). Then, are there any other operating modes such as single train operation on a multi-train process, on-line catalyst regeneration or cleaning-in-place following completion of the processing steps? Any additional modes of operation that are studied will require significantly more time, so make sure they form part of the terms of reference and are assessed in your time estimate.

It is worth spending time thinking carefully about the deviations you are going to use in continuous HAZOP because the more deviations you use, the more time each node will take (and the more bored your team may become). It is likely that you will generate 80% of your worksheet content from the 'flow' deviations (assuming you start with these in each node), so having another 20 deviations to go through — most of which will generate few hazardous scenarios — may do little but take more time and exasperate the team. We'll discuss deviations in more detail in Section 3.6.

We'll also discuss node identification and planning later in more detail in Section 3.5. At this stage it is just worth saying that if the organisation has guidelines for identifying nodes, or if you have discussed and agreed guidelines with the client, then it is worth recording them in the terms of reference. This will serve as a clear basis for the way you develop the nodes and will also act as a reference point should there be any suggestion during the study that the node policy should be changed, for example by expanding nodes to save time (not recommended by the way!).

Another aspect of the way you apply HAZOP methodology that you will want to make clear in the terms of reference is the way that you will develop and record the causes and consequences of hazardous scenarios, how you will do any risk assessment or ranking of scenarios if this is required (including how you will use a risk matrix if one is provided) and how recommendations will be developed (and possibly classified). The form of recording should be specified (proprietary software hopefully, but possibly using Microsoft Word or Excel or an equivalent). An important point here is to stress that you will use 'full recording', that is, you will ensure that all meaningful discussions are recorded, even if they do not lead to the construction of a hazardous scenario or a recommendation, and you will seek to document all causes and consequences of hazardous scenarios thoroughly. If your client does not want to use full recording and wants you to record 'by exception', that is, only when discussions lead to a significant hazardous scenario (or, perish the thought, a recommendation!) then a serious discussion needs to take place. Recording by exception, while it used to be common in the pre-computer age, is no longer considered to be recognised good practice and you should resist it. Yes, it may save some time, but it is extremely difficult to review a study that has been recorded by exception and understand what has been discussed because a lot of it will not be there. With respect to recommendations the 'what where why standalone' criteria (Section 4.1.5) should be mentioned (to stress that your recommendations will be written thoroughly in order to give maximum assistance to your client), which of course takes a little more time (although most of this should be your time at the end of each day of study, discussed in Section 7.5). It may also be helpful to think about classifying recommendations. This often takes the form of agreeing to separate out recommendations relating to P&ID or other documentation changes that may be required from more substantive recommendations that relate to risk reduction. It may be worth going further than this and thinking about classifying recommendations by subject (those requiring design changes, those requiring procedural changes, those relating to training etc.). This will very much depend on what the client wants, but it is well worth discussing because it may give you an opportunity to improve the usefulness of the final report; this is discussed in more detail in Chapter 7.

The protocol you will employ for the management of P&IDs is an important aspect of the terms of reference. It might sound excessive, but many studies have run into problems with controlling 'master' P&IDs (the actual P&IDs that are on the meeting table or meeting room wall and examined by the team) and so it is worthwhile thinking through how this will be done. You will need a scanned image of all master P&IDs to include in the final report, so these need to be marked up with the node identification carefully and neatly and kept in good condition. My preference is for the HAZOP leader to personally control a single set of master P&IDs, preferably in A0 or A1 size, stamped 'HAZOP Master' or similar and signed off as completed when all the nodes marked on them have been concluded. It may also be useful to write on them or reference any recommendations at the place in the process they apply and any other observations such as required drawing changes. And of course, make sure you have the capability to get them scanned as soon as they are completed; this may not be straightforward for large sizes!

Finally, it is useful to think through and describe how requests for further information are handled within the HAZOP meetings when the team decides it does not have all the information it requires. Meetings can easily be slowed down if additional information is being requested frequently and this will frustrate the team. Better to operate a 'parking lot' or action list whereby requests are recorded and assigned to a team member to chase up between meetings. This will save time but needs to be managed closely to avoid large backlogs developing (remember that any outstanding parking lot item at the end of the study will need to be converted into a recommendation; your client may not appreciate an even bigger list of recommendations!).

3.3.6 Decide what process safety information you need

HAZOP involves a detailed study of the process, and the only way that a sufficient level of detail can be achieved is by making sure that the relevant technical information is readily available to the HAZOP team; this information is often given the wide-ranging title of process safety information or PSI and extends to a far greater level of detail than just the P&IDs.

When discussing a possible hazardous scenario you will need to know, for example, how much pressure could be developed by a centrifugal pump running against a closed delivery and compare it to the design pressures of equipment that may be subjected to the pressure to decide if overpressure is a credible event. When considering possible safeguards you will need to check that a pressure relief device is designed to relieve the conditions of the scenario in question, that an instrumented protective function would react in a certain way or that an operating procedure tells the operator to perform a certain action in response to a process alarm (i.e. to justify safeguards by describing how they return the process to a safe state). Without supporting technical information these discussions cannot be concluded properly; you will be forced to create recommendations to find the required information and the team may become increasingly frustrated. The generation of large numbers of recommendations to find relatively straightforward technical information is a common sign of a shortfall in HAZOP preparation and an early warning, in the worst case, that you should stop the study.

Appendix 5 contains an extensive list of possible sources of PSI; the essential documents are summarised in Box 3.2.

Without this set of documents the HAZOP should not go ahead, hence the importance of listing the required documentation in the terms of reference because there is invariably a significant amount of work required to assemble it.

With the obvious exception of P&IDs, which you will need in hard copy format (preferably in large size as emphasised above), much PSI can be made available electronically and is frequently assembled on a network drive. It is important that documents are relatively easy to find or the team will become frustrated, and it is a good idea to provide a second laptop and projector so that searching and viewing documents can be done easily without disrupting the recording of the study. As an example, if a process alarm is being suggested as a safeguard it is important to be able to find the alarm set point and then view the corresponding operating procedure quickly to confirm that the alarm response is documented in a way that it will bring the process back to a safe state; otherwise, it is not a valid safeguard and should not be recorded as such. As we'll stress later in Chapter 4 (Section 4.1.4) 'Alarm PIA-123' is not a sufficient description of a safeguard: we need its set point, what the required operator action is, in which procedure, document or other place the operator action is prescribed, and how it will bring the process back to a safe state.

BOX 3.2 Essential process safety information.

- Process description and chemistry
- P&IDs 'Approved for HAZOP'
- Process Flow Diagrams
- Facility plot plan/unit layout drawings
- Equipment design conditions
- Control, alarm and trip information
- Pressure relief, flare, vent and de-pressuring information
- Operating procedures
- Previous HAZID, HAZOP or LOPA reports
- Changes to design since the last HAZOP
- Previous incident reports

3.3.7 Who needs to be involved?

Arguably more fundamental to the quality of the study are the size, composition and consistent attendance of the HAZOP team. The blend of knowledge and experience, and the healthy dynamics engendered by consistent attendance, are crucial to achieving the required depth of challenge of the design and creative discussion of possible deviations. This is a prime concern of the HAZOP leader and therefore it is vital that you play as big a part possible in assembling your team. The terms of reference are an important tool with which you can try and achieve this, in three ways:

- by specifying **core** roles, without the attendance of which HAZOP meetings will not take place;
- by specifying named persons and their deputies for these roles; and
- by specifying named persons in additional non-core roles (to try and limit the size of the team).

Typical core roles are shown in Box 3.3.

The project engineer and vendor representative roles are shown in parenthesis to indicate that they will only be needed where the HAZOP involves vendor-supplied equipment and/or relates to a project.

Box 3.3 makes the important point that the HAZOP team should be limited in size, wherever possible, to 5−7 persons. This is to assist you in your task of facilitating effectively: the larger the team, the more difficult it will be to facilitate. Depending on the nature of the plant or process to be studied, other functional engineers and technical experts such as chemists may be required but you can try to limit this involvement to an 'as required' basis if you are concerned about numbers. Some organisations employ the role of independent process engineer (IPE) to provide an additional challenge to the process design engineer. This and other aspects

---

BOX 3.3 HAZOP core team roles.
- Independent facilitator
- Process design
- Day-to-day operations
- Equipment design and condition
- Recorder
- (Project engineer if necessary)
- (Vendor representative if necessary)

of the team are discussed some more detail in Section 3.4. The key thing with respect to the terms of reference is to try to agree core roles and names; names may not always be possible to identify some time in advance, but agreeing the core role concept and named persons will help you to manage the risk of too few or too many people in the HAZOP meeting as well as the risk of inconsistent attendance.

3.3.8 Schedule and deliverables

What you really want to try to avoid once the study gets underway is any disagreement with your client or sponsor; you'll be busy enough without that. Such issues often emerge from differing expectations, most often in relation to how much time the study is taking (and therefore how much resource it is tying up). Having a shared understanding of the plan of execution (when and where the sessions will be held and over what period of time) will enable you to manage expectations by tracking and regularly reporting on progress. Other sources of friction with clients or sponsors can arise when major hazard scenarios are identified that the HAZOP team believes are insufficiently controlled or mitigated, or when significant weaknesses in operating plants or management systems are uncovered ('This unit has never worked properly since it was installed; management has never listened to us'). It is best to highlight such issues as soon as they are uncovered, rather than risk criticism ('Why did you wait to submit your report before telling me this?'). Such considerations can be set out in the terms of reference by specifying the types of issues that should be immediately reported: an example would be the identification of 'high risk' scenarios as assessed using the organisation's risk matrix; another might be the identification of a major hazard scenario that is only controlled or mitigated by human intervention.

The ultimate deliverable for the study, of course, is the final report, so it is worth thinking about how quickly you will be able to deliver the final report following the final HAZOP sessions and agreeing this in the terms of reference. Linked to this is the question of how the organisation intends to deal with the recommendations from the study. If you understand how this is to be done, for example using a company action-tracking software system, then you may be able to arrange for this system to be populated as the study progresses, or at least to develop recommendations in a format that is compatible with it. If the organisation hasn't yet thought of how it will deal with them, then you may at least help them to avoid a nasty shock by helping them to start thinking about it. Many experienced HAZOP leaders have returned to lead a repeat study on an existing facility only to find that the recommendations from the previous study were never addressed. As a contracted third party it's always a nice feeling when you are invited back, but it's really discouraging to hear that the project you put so much effort into last time has achieved nothing!

3.3.9 Report content and distribution

Finally, again on the theme of managing expectations, it is useful to specify exactly what the final report will look like: the contents of the main body of the report and the information that will be appended. It is also helpful to specify the main recipient by name, as well as the planned distribution. Knowledge of the main stakeholders can help you in the way you lead the study as well as in the way you present the final report. This is covered in detail in Chapter 7.

To conclude, agreeing terms of reference provides you with a means of clarifying the objectives and scope of the study and a tool to manage expectations and deal with emergent issues as the study progresses. It's your contract and your insurance policy, so producing and agreeing it is always well worth the effort!

3.4 The HAZOP team

So terms of reference are important in helping you to make sure the team that is brought together has the best chance of producing a thorough and effective study in terms of the blend of skills and experience in the team, its size, and the continuity of team members' involvement. Ideally, the HAZOP leader is fully involved in the planning of the study and has a degree of discretion in terms of team selection. But in reality, of course, this may not be the case. The organisation, client or sponsor may have fixed ideas of what the HAZOP team should look like, may not be willing or able to identify team members in advance of the study or may not be willing or able to commit to maintaining the same fixed team throughout its duration. Your difficulties as a leader could be further compounded if you are not part of the organisation (and therefore have less influence) or if you are brought in relatively late in the preparation phase, both of which are fairly typical for independent third-party (contracted) leaders. In this section we'll examine the roles of the different team members in more detail in order to help you navigate some of these potential difficulties and then discuss ways in which you can help to prepare the team for a 'flying start' to the study.

3.4.1 The core HAZOP roles

Essential roles in the HAZOP team — the 'core' roles that should be filled in every meeting (if you are able to agree to that stipulation) — were presented in Box 3.3; we'll consider each of these roles in turn.

Having discussed the responsibilities of the independent facilitator in detail in Chapter 2, we don't need to repeat them here, other than to stress the responsibilities as both project manager (or rather delivery manager) and meeting facilitator, the combination of which is aimed at planning and producing a thorough and effective study. Focusing on the role of facilitator, your responsibility is to get the best possible performance from the team, hence the stress on getting involved in team selection if you can so as to make the facilitation aspect as straightforward as possible. It is also important to stress here one responsibility that you should not have, and that is acting as a technical expert. The more modest and self-respecting HAZOP leaders never tire of saying this, although they are not always listened to. It is incredibly disheartening to be criticised after a study for not having understood the process well enough, but it happens. Your job as facilitator is to get the right technical information from the team, and that means asking questions like 'can you explain how this works?', even if you think you know. This is not displaying lack of technical knowledge (although obviously to some who don't understand the role of HAZOP leader it might look like this): it's just good facilitation. On the other hand there are HAZOP leaders that are more than happy to make technical suggestions and indulge in technical discussions, but this carries the risk that team members may sit back and let you develop all the suggestions or that you get involved in technical discussions to the extent that the vital facilitation tasks of monitoring progress, quality and team dynamics are neglected. This subject is discussed in more detail in Chapter 5. In summary, it's helpful to have technical knowledge and understanding of the process, but it's not your role within the team.

The technical understanding of the process is the prime responsibility of the process design role. This role is most often filled by a process (chemical) engineer. For a new design or modification the ideal candidate is the process designer and for an operating plant it is often the process engineer responsible for the facility. This person must be able to explain the design intent for each node of the study and then participate in most of the discussions relating to deviations from design intent. There is no hiding in this role! Needless to say, the role is indispensable and any short-comings in knowledge and experience (or attendance at meetings) will very quickly be exposed and will present a major threat to the effective-ness of the study. It is also worth mentioning that the process design rep-resentative can feel very exposed in HAZOP; individuals can sometimes feel like they are being tried in court or that their designs are being criti-cised. The HAZOP leader must be keenly aware of this possibility, watch out for signs of discomfort or stress on the part of the process design rep-resentative and provide appropriate support and intervention if necessary.

Experience of day-to-day operations is relatively straightforward in the context of a study on an existing facility, in that the experience can be provided by a process operator or perhaps a supervisor. From the point of view of selecting such a person, considerations will be the breadth and depth of experience and, very importantly, the willingness and capability of the representative(s) to participate in what may be the unfamiliar and − potentially uncomfortable, or even intimidating − environment of a meeting room, compounded by the feeling that they may perceive them-selves as less valuable than other team members on the basis of rank or seniority. Again, this signals the need for careful monitoring and support from the HAZOP leader. Another aspect of the operations role that often presents difficulties is continuity of attendance, given that operations are invariably shift-based and shift teams rarely carry spare berths and, in some sectors, are rotation-based, working for example 28 days on−28 days off on some offshore installations. The consequence of this is that the identity of operations representatives may change, which always brings with it potential disruptions to group dynamics. This is not just in terms of the adjustment to the group dynamic often brought about by a new face, but the new face can sometimes bring with it a very different view of how the process does or should operate! You might experience this when, reviewing the previous day's worksheet at the start of the next day, you hear an exclamation from the new operations representative of, 'Who said that we do it like that? That's rubbish!'.

For the study of a new project, it is often necessary to select an opera-tions representative from another facility that employs the same or similar technology (relatively straightforward, for example, for a new offshore production facility), but for a new facility employing new technology it may not be possible to have direct experience of the operating the technology in question. In this case it is still important to have a representative with plenty of experience of day-to-day operations of processes to bring the experience and perspective of what it is like to actually control and operate a live process 'hands on' on a routine basis.

The third important technical role requires knowledge and experience of the equipment design and condition for the process in question. HAZOP is sometimes described as 'findings ways to break the process' and, in this sense, it does consist of uncovering and exploring scenarios in which design conditions are breached and equipment is exposed to conditions that may cause it to fail. It therefore follows that the team should include a representative that can help to assess the likelihood of equipment failure and the resultant loss of containment, which requires knowledge of the design conditions of the equipment (design pressures, temperatures etc.). For projects this is likely to be an engineer with experience of mechanical design. In the case of studies on operating plant, knowledge of the actual condition of the equipment (age, maintenance, inspection history and any perceived vulnerabilities such as susceptibility to corrosion or erosion for example) is necessary, and this is often provided by the engineer responsible for maintenance or asset integrity.

Last, but by no means least, among the core roles is that of the recorder (often called the 'scribe' but to me this term devalues the role by conjuring up the image of ancient slaves and tablets of stone, so I'll use recorder from now on). Historically this role has been undervalued, and probably still is in some organisations, but these days it is widely accepted that a competent recorder is essential to achieving an effective HAZOP study and, conversely, that a less than competent recorder can seriously disrupt a HAZOP study. A typical set of responsibilities for the recorder are shown in Box 3.4.

The prime responsibility of the recorder is obviously to generate — in real time and live on screen in front of the team — an accurate, comprehensive and readable record of the discussions of the team in the form of the HAZOP worksheets. That sounds straightforward but, of course, it isn't to all but the best recorders, which is the reason for the second bullet point in Box 3.4. Anticipation is presented here as a responsibility although, in fact, it is more a skill because the way in which the recorder performs their role can enhance the smooth running of a study or disrupt it. Concentration, careful listening and comprehension will enable a good recorder to choose the right moment to start writing and help them to produce an accurate, succinct summary of the discussion quickly and

> ### BOX 3.4 Responsibilities of the HAZOP recorder.
> - Summarise the discussions of the team
> - Anticipate to maintain the flow of the meeting
> - Ensure completeness of FULL RECORDING
> - Ensure correct spelling and grammar
> - Ensure comprehensibility for the reader
> - Ensure that cross-referencing is accurate (avoid if possible)
> - Develop a glossary
> - Update the worksheets with parking lot responses; removal of parking lot and other flags

efficiently. When this is going well, it gives the team confidence and helps the study to flow quickly; when the recorder cannot keep pace or records discussions in a way that the team doesn't agree with or can't understand, the study will suffer frequent interruptions and the team (and you yourself!) will become frustrated.

Given the importance of the role, and the way that you will need to work closely as a team with the recorder within and outside of the HAZOP meetings, careful selection of the recorder is critical. As I said previously, in the past this has not always been given sufficient consideration, the recorder role being seen sometimes as a 'necessary evil' and sometimes even the need for a dedicated recorder being challenged as an unnecessary additional cost. Most HAZOP leaders have been forced, at some time or other, to facilitate and record at the same time. It doesn't work on anything but a temporary, 'needs must' basis, because the quality of both the facilitation and recording suffer. Facilitation and recording are two fundamentally different tasks, each of which competes for attention and requires the brain to work in a different way. So as HAZOP leaders we should be steadfast in requiring a dedicated recorder if a study is to be conducted according to recognised good practice. Having secured the role of a dedicated recorder, we should then assess carefully the capabilities of candidates for the role, should we be fortunate enough to play a part in their selection.

The key attributes of a good recorder are strong concentration (which implies motivation), good listening skills, good literacy skills, a technical background that enables discussions to be understood and summarised, and that all-important sense of anticipation and timing. In my experience these attributes are most often — although not always — found in young (and enthusiastic) engineering graduates, in the early years of their industrial careers, who are keen to expand their technical competence through the experience of being part of a team exploring the design of the processes and hearing experienced engineers discuss the process, as well as take on board the technique of HAZOP and the insight into the world of process safety that it provides. Motivation is important here because it is unlikely that a good recorder will wish to stay in the role for an extended period of time (they will eventually get bored), so the role of recorder is often prescribed as one (small) part of a young engineer's competence development programme.

It is worth noting that although typing skill would appear to be an obvious requirement for a recorder, where this has been used as the only selection criterion (in the form of a dictation test) it has failed spectacularly; attention, listening and anticipation are more important. As for using a non-technically-trained person as a recorder, don't even think about it. It's been tried as a way to save money but there is no evidence that it has ever succeeded.

Finally, if you as HAZOP leader are presented with a recorder to work with, without prior knowledge of their skills in this role, then it is important that you begin a process of mentoring and coaching as soon as possible. We'll discuss this further in Section 3.4.5.

The five roles described so far constitute the minimum requirement for a competent HAZOP team, hence their designation as 'core'. Box 3.3 presents two other roles that may also need to be considered as 'core' depending on the subject of the study. In a new capital project a project engineer may wish to be present throughout the study if, for example, the HAZOP is viewed as a critical activity from a timing point of view or there are concerns about the quality of the process design (or, more cynically, the project engineer wants to be present when decisions with potential cost implications are made). In my experience, I would prefer not to have a project engineer present, for the simple reason that their mind is likely to be focused on issues that could potentially impact on project cost and programme rather than engaging as part of a team to examine the process design in a creative way. Again, cynical you may say, but borne out by experience. I've experienced several studies where the project engineer has attended the opening meeting and announced that the project is late and over-budget, so they don't expect any major changes resulting from HAZOP. Not the start you want to motivate your team!

Another important core role is required, however, when the HAZOP involves the study of vendor-supplied equipment. In this case (if it is really necessary to study standardised equipment often in use across the industry) then, for the process design role we need a team member that knows in detail the design of the equipment and this is almost certainly has to be an engineer from the vendor organisation. The involvement of such an engineer should be restricted to the part of the study involving their equipment and its interfaces with the main process (they will not have any interest in anything else and will become a distraction). There are two other warnings that come with the involvement of vendor representatives. The first is that the vendor engineer is always likely to be defensive in respect of any possible changes that the HAZOP team might suggest that involves their equipment; this is perfectly understandable and is often expressed with frustration as, 'We supply thousands of identical units to suppliers all over the world every year, but it's only this organisation (or project) that seems to think that our equipment should be modified'. The second warning is related to the first, and that is the vendor engineer will often arrive with a colleague, invariably a commercial representative who will play no role at all in the study until a potential modification to the vendor equipment is suggested; they will then move in to try and avert this by way of thinly veiled threats relating to the additional costs and delays that will be incurred in supplying 'bespoke' modified equipment. Obviously, a second vendor representative is another team member and therefore another person for you to worry about, so best to avoid if this if at all possible.

3.4.2 Additional roles (if you must)

With our core roles we already have a team of between 5 and 7. Many HAZOP leaders would say that this is the 'right' size of team from a knowledge and experience perspective, and also from a facilitation perspective. Imagine yourself in the facilitator role with any more than 7: do you think you could facilitate the discussion with any more around the table, while at the same time monitoring the dynamics of the group. . .is everyone involved, is everyone looking happy and engaged, what is the body language saying?

Many HAZOP leaders start to get uncomfortable when the size of the team is greater than this, but unfortunately sometimes you will need additional knowledge and experience in the room. This is a function of the nature of the process you are studying. For example, if the process involves complex reaction chemistry, as it often does in the chemical and pharmaceutical sectors, then a process chemist is often required. If the process involves complex control systems or automation schemes, for example programmable logic controlled (PLC) processes, then a control and instrumentation (C&I) or automation engineer may be required. Or if the process involves specialist machinery such as compressors or fired equipment, then a mechanical or machines engineer may be required, and so on for other engineering functions. An important question to consider in the case that you identify a need for additional expertise is whether you need these additional people in the room all the time? If you don't then it may be possible to arrange the meeting sessions so that the processes requiring additional input are grouped or sequenced such that the additional team member is only required for certain sessions or on certain days. Alternatively, it may be possible to have the additional team member(s) 'on call' remotely and summoned when needed. In this way you keep the numbers down for as much time as possible and you also avoid the potential disruption of additional team members sitting in the meeting getting bored when their specialist subject is not being discussed. Of course, you may be concerned about possibly offending these people by inviting them only on a part-time basis, but you are much more likely to get their thanks!

3.4.3 Independent process engineer

A significant number of organisations have, in recent years, introduced the role of Independent Process Engineer (IPE) as a core team member; that is, a process engineer who is independent of the organisation that has designed the process, often hired on a third party, contracted basis. The concept here is to provide some additional challenge to the internal process engineer from someone with significant process engineering knowledge and experience; this is particularly relevant if the internal process engineer is less experienced and especially if they comprise the only process engineering expertise in the team. The concept has some logic and merit, but I have little experience of it working well and plenty of experience of it introducing additional problems, notwithstanding the issue of adding another member to the team. One type of problem is dysfunctional interactions between the two arising from a 'clash of egos', aggressive challenging by the IPE or lack of respect on the part of the internal process engineer arising from the IPE's lesser knowledge of the process (or vice versa!). Another type of problem is the IPE's behaviour in general and the way it affects the team. The IPE is invariably a (well-paid) contractor who may see the role as relatively uninteresting (but lucrative) and consequently doesn't take it as seriously as the other team members, which will quickly be noticed by them. The answer here, if you like the idea of an IPE or are required to use one, is to select that person carefully, looking not just for knowledge and experience but for motivation and interpersonal skills (the ability to question in a constructive fashion, empathy and the capacity to 'see the other point of view').

3.4.4 Safety manager

Some organisations require a safety manager or a process safety technical authority (where this concept is employed) to be part of the HAZOP team. There is obvious logic to this, in that HAZOP is very much a part of the process safety arena. A safety manager who does not specialise in process safety is unlikely to add much to the skills and experience of the team and becomes a further addition to the facilitation load. A process safety manager or technical authority is likely to bring additional knowledge and experience, but is it necessary? In my experience it can be helpful if it brings understanding of the facility in relation to major hazards regulations (e.g. a site's Seveso III report or safety case and the scenarios contained within it), but outside of this, it is less likely to add additional value, and therefore best to have this role 'on call' or remotely contactable if and when required.

3.4.5 Preparing the team

Having a named team in place well in advance of the study is a significant achievement in itself and it is tempting, once the terms of reference are agreed, to divert your attention to your other projects or responsibilities. However, as we'll see later there is a lot of preparatory work to do in the run up to the study and it is worth considering the team as part of this, especially if you have never met them before.

Once the study starts, the 'clock is ticking' in terms of your agreed schedule and you will want to get the study off to a flying start. This means getting the team working effectively as early as possible. If you can arrange to meet the team before the study or, failing that, meet each of the team members individually, then you can learn a little about their backgrounds and experience and this will start the process of team development; it will also signal to the team your enthusiasm for the study and the value you place in them as members of the team! Of course, this is not always possible, and so you can try to at least call each member and introduce yourself, ask for a copy of their CV and perhaps ask if they have any concerns about the study and its preparation.

If you are fortunate enough to be able to meet as a team, then you can share thoughts and concerns about the study, discuss the progress of preparations, find out whether any team members require training in the HAZOP methodology and introduce the subject of 'ground rules' — the rules that you want to develop with the team that define the agreed desirable (and undesirable) behaviours that you will all promise to adhere to (or avoid). Doing this properly with the team (rather than arriving on the first day with a list you have prepared yourself) is far more powerful than trying to impose behaviours, but obviously takes time. So if you can do some of this work upfront you will be able to start the HAZOP analysis earlier.

Arranging to meet team members can be very difficult, especially for a third-party, contracted HAZOP leader. It's not the end of the world if you don't succeed, but it is worth making every effort to try and meet at least one of the team members — the recorder — if you have never met. You are going to have to work as a close-knit pair throughout the study and, in most cases, you will be acting as a mentor and coach, so the sooner you start the relationship the better. It is very important that you have a shared understanding of recording requirements set out in the terms of reference (full recording, use of tag numbers etc. as shown in Box 3.4) but also communicate your own personal expectations of the recorder's behaviour. Examples of such expectations might be: encouraging the recorder not to wait for you to dictate to them, to ask if they don't understand something or speak up if they are feeling overwhelmed with multiple simultaneous voices; it is so important that the recorder feels that you will support them if they get into difficulties.

In Chapter 5 we'll cover the subject of facilitation, and this will obviously bring us back to the subject of the team and the individual behaviours of its members. In the meantime, we'll move on in this chapter to cover two technical aspects of the application of HAZOP — node identification and deviations — which will lead us into estimating how long the HAZOP study will take, and then into the planning in preparation for the event.

3.5 Node selection

Node selection (node definition or identification) is an important subject for two main reasons. First, it is the most accurate source for estimating the amount of time that will be required to undertake the HAZOP study. And second, the size and relative complexity of the nodes can have a significant impact on the ease of facilitation and the speed and effectiveness of the study.

Given the importance of the activity, the HAZOP leader should be fully involved, preferably working alongside the process engineer who will be participating in the study, who can provide technical explanations to support you in your decisions. It is important to emphasise that node selection takes time, and the availability of the process engineer to assist you may well be an issue, so plan for it if you can.

If nodes have already been defined by the time your join the project, make sure you are comfortable with how this has been done and how it relates to the estimation of the time required to complete the study. You will be held accountable for meeting the schedule and, once you start the study, changing node definitions can create administrative difficulties in terms of marked-up P&IDs and pre-defined node descriptions.

The implication here is that node selection is carried out at the earliest opportunity so that you can use the total number of nodes as the basis of the estimate for the time required to complete the study, which will then determine the schedule to be set out in the terms of reference. As explained earlier, it is helpful to include a list of P&IDs in support of the definition of the scope of the study and, if you have the set of P&IDs, then you can undertake the node selection and time estimation. We'll discuss time estimation in Section 3.7.

Sometimes things don't always go to plan and you may get access to the P&IDS later than you would like. In this case, make sure the time estimate is revisited once you've defined the nodes to make sure you are comfortable with achieving it; alert the sponsor or client sooner rather than later if you believe the time estimate is too short. Never leave node selection to the study meetings themselves. Although some HAZOP leaders believe that 'democratically' identifying nodes with the team is a good way to bring the team together, this process is incredibly time-consuming and boring for the team, and can send a very clear message that you have not prepared!

3.5.1 Node selection for a continuous process

There are no precise rules for node selection and yet selecting an appropriate node size and guiding the team through the node is crucial for the success of the HAZOP study. Nodes that are very small, such as sections of a single process line, often lead to longer study times because each deviation must be recorded more times (there is also likely to be a lot of repetition). Large nodes, which might include multiple process lines and equipment items, may generate confusion in the application of the deviations ('which line are we looking at?') and, if not properly managed, present a risk of hazards being overlooked. Factors that can influence the decision on the size and complexity of a node include the experience of leader and team, the severity of the hazards of the process, and the complexity of the control system.

To ensure that the design intention of each node can be easily and clearly understood, nodes should be selected by function (examples being such as 'transfer', 'cool', 'heat', 'react' etc.), as shown in Fig. 3.2.

The following criteria should be considered in selecting the appropriate transition to the next node:

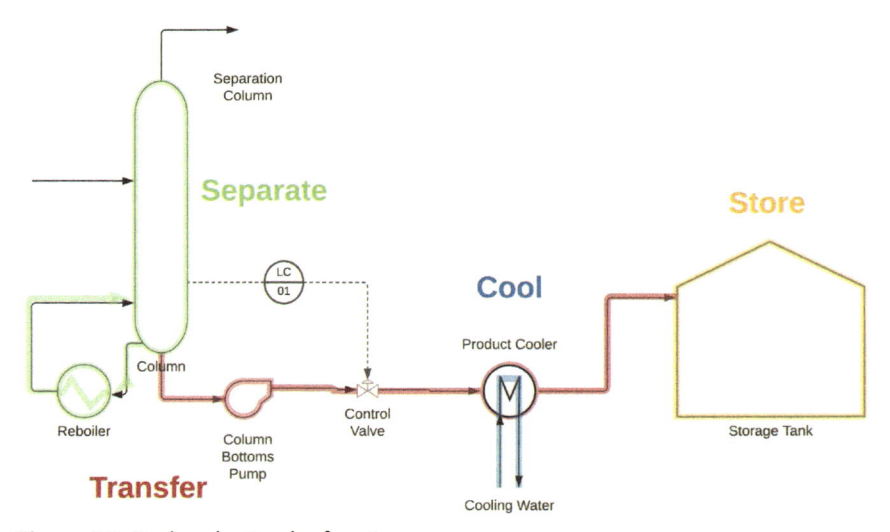

Figure 3.2 Node selection by function.

- a change in design intent;
- a change in state (e.g. from liquid to vapour) or composition;
- a major piece of equipment.

The usual process for node selection is to work through the process in the direction of flow using the PFD for overall guidance. At its simplest, nodes can be formed from the process lines connecting each major vessel on the PFD, with the major vessels forming their own nodes and minor equipment such as pumps between major vessels included in the same node as the process line.

If nodes are selected that contain multiple and/or branching lines, this could create confusion; for example, if the deviation is 'more flow', there may be confusion over which line is being discussed at any one time. In this case the leader must ensure that team members are thinking about the same line, by systematically guiding the team in reviewing one line or branch at a time (there is extra stress for you too, in systematically working through it). Rather than run the risk of confusion, better to follow the words of process safety guru the late Trevor Kletz, '...the guide word should apply uniformly throughout the node' [4], which suggests the choice of single lines as the best option (balanced against the disadvantage of having more small nodes).

Guidance on node selection in relation to typical process equipment such as vessels, process lines, pumps and heat exchangers is provided in Appendix 6.

3.5.2 Nodes for complex systems

Complex systems tend to cover multiple P&IDs and multiple connections. Examples are flare/vent headers, drainage systems, fuel gas (and liquid fuel) systems, inert gas, instrument air systems, distribution manifolds and lines, process water, cooling water, fire ring mains and chemical dosing systems (ring-main type systems).

Complex systems need to be studied in multiple parts. The connections into these systems from the process need to be studied first as part of the relevant process nodes, the nodes ending at the isolation valve upstream of the connection point (if there is one).

For a complex collection system such as a flare or drainage system, select one node for each separate header or manifold as far as the collection vessel, making sure that each node starts at all the upstream connection points. In the case of a flare system the knock-out drum and flare

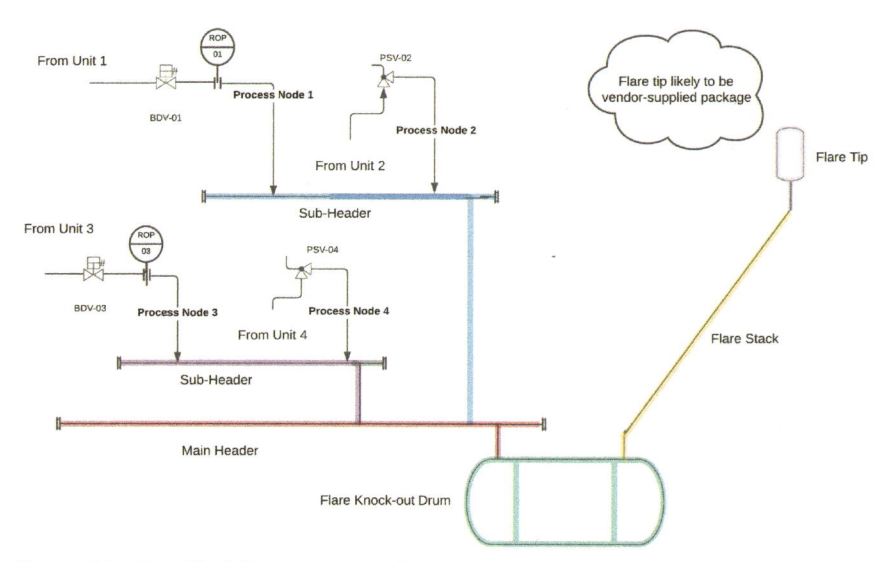

Figure 3.3 Simplified flare system nodes.

itself should be identified as separate nodes. It is useful to develop a list of all process flow conditions into flare/vent header system — continuous and intermittent flows, process conditions and expected flow rates and pressure conditions. Fig. 3.3 illustrates the principles of node selection in a simplified example of a flare system, where the nodes are shown in colour.

Complex supply systems (chemical dosing, inert gas, instrument air, cooling water, process water) can be divided into nodes using the same principles: process nodes extending to the interfaces with the complex system; a node for each supply main, header or manifold in the supply network (meeting each relevant process connection); a node for each return system (in the case of cooling water) and further nodes for equipment such as air or inert gas receivers and then for supply pipework from the source equipment such as compressor or dosing skid. Fig. 3.4 shows a possible node selection for a simplified cooling water system.

3.5.3 Different operating modes

If a node has more than one operating mode, for example normal production and in-situ molecular sieve regeneration, then each operating mode or condition should be considered as a separate node. It may be that the operating mode is a sequential operation, for example

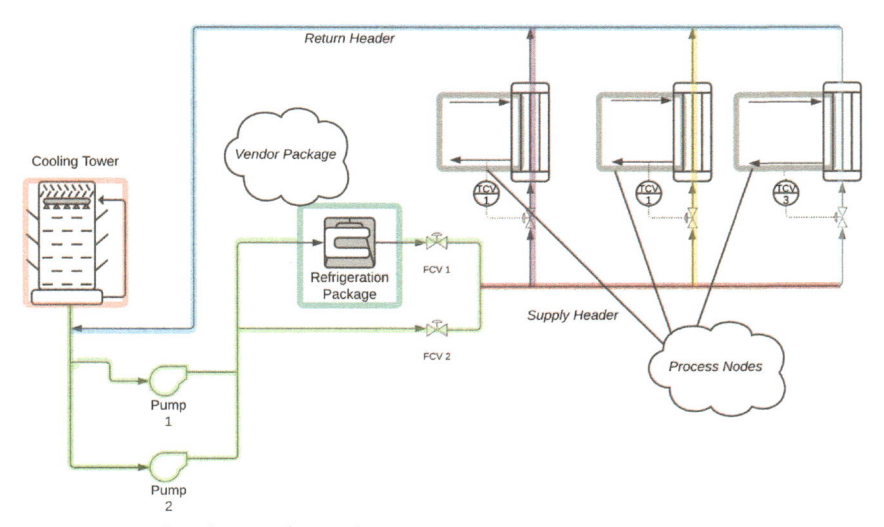

Figure 3.4 Node selection for cooling water system.

cleaning-in-place, that would benefit from procedural HAZOP rather than continuous.

3.5.4 Parallel trains

Parallel trains may be reviewed independently or one train may be reviewed and the next train may be reviewed 'by difference' based on the first. If the latter approach is taken, the trains must be compared in detail to ensure that any differences in control, instrumentation, piping arrangement, and equipment design is identified and considered. Beware HAZOP by difference! Parallel trains are rarely identical and are best studied by the same team that studied the first train, working through the HAZOP worksheets from the first train line-by-line looking for differences. There may also be a need to identify another mode of operation if single train operation differs from the normal two-train operation.

3.5.5 Study boundaries or interfaces

Rather than ignore everything upstream or downstream of the limits of the study, it is appropriate to apply deviations upstream outside of the node, although precise causes (upstream) or consequences downstream may be difficult to evaluate if the nature of the operations outside of the study are not fully understood. An example of this would be where the operation upstream of the study may be the supply of a key raw material or utility from a third party, so that the 'no flow' deviation elicits the cause 'loss of upstream supply'; the precise cause may not be identifiable, but if there are concerns about potential downstream consequences, then recommendations or a list of concerns can be raised to investigate further with the supplier outside of the study.

3.5.6 Node selection for batch and procedural HAZOP

For a batch process or procedural HAZOP, nodes represent the main process or procedural steps. To allocate a separate node to every single process step is likely to result in a large number of nodes and a significant amount of repetition. In batch processes, it is best to group the detailed procedural (or PLC-driven) steps into node such as:

1. addition of solvent
2. addition of reactant A
3. additional of reactant B
4. reaction
5. cooling
6. product transfer

For procedural HAZOP, conducting a hierarchical task analysis [5] on the operation can help to identify the safety-critical parts, which can then be subjected to HAZOP. This focuses the HAZOP effort on the most important aspects of the operation and avoids excessive effort. Fig. 3.5 shows a hierarchical task analysis for a chlorine unloading operation.

This analysis would suggest the application of procedural HAZOP to the 'connect transfer lines' step since this is the step in which loss of containment could be considered to be most likely.

3.6 Deviations from design intent

As has already been said, it is worth spending time thinking carefully about the deviations you are going to use because the more deviations you use, the more time each node will take and the higher the risk that your team may get bored as you plod through unproductive deviations (it's a vicious circle: the more tedious it is, the less contributions there are).

Some organisations have a standard list of deviations that are required to be used, in which case make sure you know about this before you estimate how long the study is going to take. I have regularly experienced studies in which the organisation has a 'standard' list of perhaps 20 or

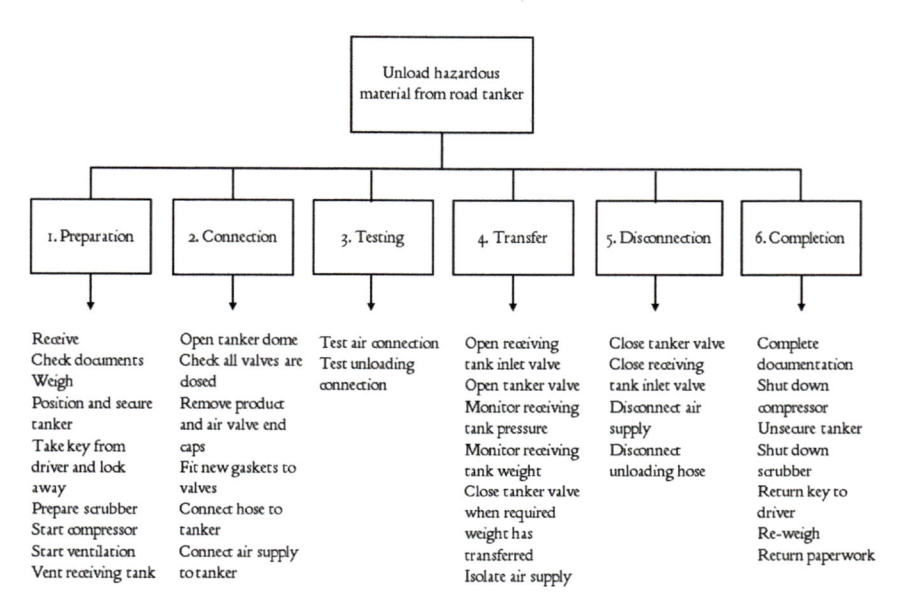

Figure 3.5 Chlorine unloading hierarchical task analysis.

more deviations (including arguably spurious ones such as 'start-up', 'shut-down' and 'maintenance') and was once told of a study on a pharmaceutical process that used 65 deviations for each node! Having to deal with a long 'tail' of deviations can not only exasperate the team but can create an impression of slow progress. If you are fortunate enough to be empowered to select or recommend a set of deviations, then start with the basic most commonly used deviations.

3.6.1 Continuous HAZOP

The 14 most commonly used combinations of process parameters and guide words that are combined to form deviations used in continuous HAZOP are shown in Table 3.1.

These deviations are termed 'principal' because they are used in just about every study; a study that does not employ these deviations would not be considered as being aligned with recognised good practice. The 'level' deviation is sometimes not relevant, for example in a node consisting of only pipework, but it is usually part of the standard set in that most processes at some point have equipment in which there may be a level of fluid. It might be argued that a process fluid is not expected to change in composition when a chemical reaction or separation is not part of the

Table 3.1 The most used deviations in continuous HAZOP.

Principal deviations for continuous process plant						
Process parameters	Guide word					
	No	Low	High	Reverse	Misdirected	As well as/other than
Flow	No flow	Less flow	More flow	Reverse flow	Misdirected flow	As well as/other than flow
Temperature	—	Low temperature	High temperature	—	—	—
Pressure	—	Low pressure	High pressure	—	—	—
Level	—	Low level	High level	—	—	—
Composition	—	Low composition	High composition	—	—	—

process, but this deviation can be used to identify hazardous deviations relating to contamination or degradation.

Starting with the deviations in Table 3.1, we then need to think about whether there are any additional deviations that might help to provide additional focus on important hazards relating to the process. Examples relating to specific technologies might include using the deviation 'more re-liquefaction' in a study involving the use of chlorine, 'more hydrate formation' in an oil and gas production study or 'more dead-spaces' in a study involving a chemical with potential to polymerise explosively like hydrogen cyanide. Use of the deviation 'corrosion' (it should really be termed 'more corrosion' or 'accelerated corrosion') could be useful in the re-HAZOP of an ageing process or a process known to suffer from corrosion. The use of such specific guide words should be the exception rather than the rule; a well-facilitated team should be able to develop appropriate hazardous scenarios using the set of most commonly used deviations. An example of an exception to this would be HAZOP of a process involving reactive chemistry, particularly exothermic reactions with the potential for uncontrolled 'runaway'.

Appendix 7 contains a large set of deviations that have been used across many organisations, including the most commonly used as well as the reactive chemistry-specific ones, and others like corrosion mentioned above. It appears that over many years new deviations have steadily crept into use, some useful but others less so. Consider the deviations 'equipment' and 'instrumentation': what do they mean?; are they really deviations from design intent? 'Maintenance' and 'training' again are not deviations and have little meaning when adequate maintenance and training are underlying assumptions in every HAZOP study. And finally, 'safety' and 'environment' are not deviations from design intent but simply 'catch-alls' that rarely work as such because they are unspecific and are always used at the very end of the node, when the team is ready to move on. As a HAZOP leader in a study with these deviations, you would have to interpret them in some way to explain what you expect of team members. Leaders often end up saying something like, 'Can anyone think of any possible maintenance issues in this node?' As I said, it's not a deviation from design intent (and moreover adequate maintenance is a basic assumption in HAZOP). Perhaps you could interpret it more specifically as 'Can anyone see any examples of where our site's isolation standards could not be applied?' This is more like a deviation from design intent (the intent being to design a process in which the site's standards can be applied). If you are concerned about the interpretations of any deviations you are asked to include, seek clarification from your client or sponsor and carefully explain the interpretation that you are going to apply or you will run the risk of team members making their own interpretations.

The deviations 'start-up', 'shutdown' and 'decommissioning' deserve special consideration: are they deviations from design intent? No, they are more like 'catch-all' prompts for the question, 'What could go wrong during these operating modes?' This might prompt members of the team to identify some possible hazards, but it is unlikely to elicit a full set of hazards since the deviation doesn't prompt a structured examination of the start-up or shutdown process. If these operating modes are important then it is better to subject the relevant operations to procedural HAZOP than subject every node to these same unspecific questions, especially when the questions are raised towards the end of the node when the team are ready to move on. Remember that, although most continuous processes are typically only in 'non-routine' modes of operation like start-up, shutdown and on-line maintenance for around 5% of the time, it has been estimated that 70% of major process accidents relating to failure of mechanical integrity occur during these operations [6,7], mainly on account of the increased human intervention required. If ever there was a case for the more widespread use of procedural HAZOP for non-routine operations then this is it!

Overall, the message is: restrict your use of deviations to those which you expect to add value (remembering the rule of thumb among HAZOP leaders that you'll likely get 80% of your hazardous scenarios from the 'flow' deviations if you start with them); avoid meaningless 'pseudo-deviations' like the ones discussed above that definitely will not help, unless you can interpret them as specific deviations; and promote the use of procedural HAZOP for non-routine operating modes.

3.6.2 Batch and procedural HAZOP

For batch and procedural HAZOP the 'standard' set of deviations are listed with their meanings and some examples of each in Table 3.2.

These are fairly standard for a stepwise process although, as mentioned already, they may be augmented by additional deviations from Appendix 7 that are reactive chemistry-specific.

Once we know how many nodes we have to deal with, and how we are going to apply the methodology to them, in terms of HAZOP style

Table 3.2 Standard batch and procedural HAZOP deviations.

Guide word	Meaning	Example of deviation
NO (NOT or NONE)	The activity is not carried out or ceases	No flow in pipe, no reactant charged to process, batch not cooled, check omitted, no catalyst, etc., valve not opened
MORE OF	A quantitative increase in an activity	More (higher, longer) quantity, flow, temperature, pressure, batch, concentration, time
LESS OF	A quantitative decrease in an activity	Less (lower, shorter) quantity, flow, temperature, pressure, batch, concentration, time
MORE THAN (AS WELL AS)	A further activity occurs in addition to the original activity	Impurities present, extra phase (solid or gas in liquid phase), additional (unplanned) process operation such as additional valve opened
PART OF	Only some of the design intention is achieved	Valve sticks part-opened or closed
REVERSE	The logical opposite of the design intention occurs	Vessel heated rather than cooled or vice versa
OTHER (THAN)	Complete substitution. Another activity takes place	Operation on wrong equipment

and deviations, we can make an estimate of how long the study is going to take.

3.7 How much time do you need?

There are no widely accepted rules for estimating the time required to undertake a HAZOP study, but a number of relatively simple methods have been developed as well as some more complex ones. Before we review some of these, it is worth considering some of the factors that will influence the time required. These are shown in Table 3.3; most of them have been mentioned or alluded to already.

There are no widely accepted rules for estimating the time required to undertake a HAZOP study, but a number of relatively simple methods have been developed as well as some more complex ones. Before we review some of these, it is worth considering some of the factors that will

Table 3.3 Factors affecting HAZOP length and pace.

Process	Methodology
Scale of process (study scope)	Batch or continuous
Nature of process hazards	Number of operating modes
Complexity of process	Number of deviations
Novelty of technology	Style of recording
Quality and completeness of design	Node complexity
Information	**Team**
Quality of P&IDs	Experience
Extent of process safety information	Size and continuity
Accessibility of information	Leader and recorder experience and quality
	Work pattern

influence the time required. These are shown in Table 3.3; most of them have been mentioned or alluded to already.

You can see that a whole range of different inter-related factors will have an influence on the length of the study. However, there are no methods for estimating how much time will be needed that take into account all, or indeed many, of these factors. The simple methods that do exist are based on either number of P&IDs, number of nodes or number of main plant items (MPI). Once you have made an estimate using one or more of the simple methods, it is up to you to factor in further time based on the other factors in Table 3.3, or additional factors that you think could be significant. Your own experience as a HAZOP leader is a very important consideration here and most studies of time estimation methods acknowledge this [8].

Most estimation methods — more like rules of thumb — that have been published relate to studies of much longer than 1 week duration. This is because the first few days of a study are likely to be relatively slow due to the need for introductory activities and the team (especially a new team) to develop a shared way of working. Once that dynamic is established then progress accelerates to a consistent pace, in a manner similar to the concept of a learning curve. Because of this initial 'start-up' effect (discussed in greater depth in Chapter 5), the duration of smaller studies is inherently more difficult to estimate, whereas for longer studies, if the pace is monitored and progress slows then remedial actions may be available to speed up the study again and achieve the estimated duration. For any study, it is sensible to assume that the first node will not be completed on the first day of the study or, more pessimistically, that no significant progress at all will be made on the first day! As the HAZOP leader, building in a pessimistic assumption for the first day will give you space to spend the right amount of time on the introductory activities like setting the scene and developing ground rules. Adequate time spent up front like this almost always paid back as faster team development once the study starts.

It is important to include preparation and reporting time in the estimated duration of the study, particularly if significant resources are required, as they are for larger studies: assembling PSI, selecting nodes and pre-populating the HAZOP worksheets with node descriptions are all very time-consuming (even more so without dedicated assistance from the organisation). Allowing adequate time for preparation can significantly reduce the time required for the HAZOP meetings, and therefore the length of time for which team members are required. We'll discuss this again once we've looked at some of the estimation methods.

3.7.1 Rules of thumb based on P&IDs and nodes

The rules of thumb described in this section have been developed from experience of relatively large studies, mainly in the oil and gas and petrochemicals sectors, conducted by European-based organisations operating in Europe, the Middle East and Asia; they have not been formally published, but rather shared between organisations and practitioners. The simplest rule of thumb for time estimation for a continuous HAZOP study is based on the number of P&IDs. You may be forced to use this crude approach for an initial estimate if you are not able to conduct node selection before signing off the terms of reference. This rule of thumb is 1−2 P&IDs per day based on 6 hours of meeting time and does not include preparation or reporting time (the time required to prepare the final report once the meetings are concluded). It assumes that the P&IDs are of good quality, are not complex and are up-to-date; the wide range reflects the fact that P&IDs can differ considerably in complexity (density).

A more useful and accurate rule is based on the number of nodes. For a continuous study this is 2−3 nodes per day (again, 6 hours of meeting per day); for a batch or procedural study the rule is 1−2 steps per day, again, not including preparation or reporting time. These simple rules are summarised in Table 3.4.

Table 3.4 Simple rules of thumb for HAZOP time estimation.

Style of HAZOP	Rate of progress
Continuous	1−2 P&IDs per day
	2−3 nodes per day
Batch/procedural	1−2 steps per day

HAZOP Team Meeting Time Vs. No. of P&IDs

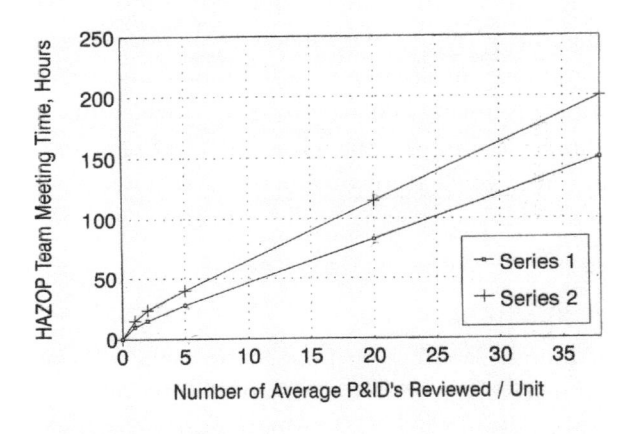

Series 1: Experienced Team
Series 2: Average Team

Figure 3.6 US example of HAZOP time estimation [9].

It is important to remember to include additional time where different operating modes will be studied using Procedural HAZOP, that is, to add on Procedural HAZOP nodes for the study of start-up and shutdown for example.

Estimated time requirements published in the USA have suggested 4−6 hours per P&ID as shown in Fig. 3.6 [9], not including preparation or reporting time.

This is broadly in line with the 1−2 P&IDs per day from European experience. However other sources in the USA have suggested considerably higher rates of progress of 2−6 nodes per day and 2 hours per node (or 3 nodes per day) [10]. I would not advocate performing HAZOP at an average rate of 3−6 nodes per day, although there may be occasional days where 3 or 4 simpler nodes can be covered in this time.

BOX 3.5 Two UK HAZOP time estimates based on main plant items [11].

Time required = [no. of MPI × 3 hours per MPI] + 3 hours

Time required = [[total number of connections to MPIs]

$\qquad$ × [20 minutes per connection]] + 3 hours

3.7.2 Rules of thumb based on equipment

Other estimating methods have been published based on the number of MPI. Two examples developed in the UK in the 1980s are based on the total number of MPI are shown in Box 3.5 [11].

The 3 hours in each of these equations represents the non-HAZOP time at the start of the study meetings, but they evidently do not include the reporting time.

A further example from North America quotes 2.5–3.5 hours per main plant item [12], which is similar to the first example in Box 3.5. It also quotes that the total HAZOP time required can be estimated as 1% of the design time (presumably the time taken to design the process, although no further details are given such as whether it relates to detailed design time or overall design time).

In practice, it may be worth estimating the time required by a number of different methods and then either selecting an average or worst case from the estimates. My approach has always been to estimate using the P&ID per day and node per day rules and take the more pessimistic estimate; however, the other estimation formulae may work for you.

3.7.3 More complex methods

Box 3.6 summarises a method published in the 1990s [8]; further guidance on the choice of the various factors are given in the reference.

For the study of a chemical process represented by 20 simple P&IDS the tool estimates a study time of 20 hours (3–4 days) for an experienced leader; for a process represented by 20 complex P&IDs it estimates a study time of 160 hours (27 days), again for an experienced leader. By comparison, using rates of 2 simple P&IDs per day and 1 complex P&ID per day for a 20 P&ID study gives estimates of 10 and 20 days, respectively.

BOX 3.6 A more complex time estimation method [8].

The total time required in hours (T_{total}) is the sum of the time required for preparation (T_{prep}), study meetings (T_{study}), preparation of the final report (T_{report}) and delays during the study (T_{delay}).

$$T_{total} = T_{prep} + T_{study} + T_{report} + T_{delay}$$

$$T_{prep} = 1.5(X_1 + 2*X_2 + 3*X_3 + 4*X_4)$$

where X_1, X_2, X_3 and X_4 are the numbers of P&IDs of different complexities (1 simplest, 4 most complex).

$$T_{study} = K*L_{eff}*(C_1*X_1 + C_2*X_2 + C_3*X_3 + C_4*X_4)$$

where $K = 1$ for chemical processes and 2 for petrochemical processes, C is a coefficient between 1 and 8 for different degrees of P&ID complexity and L_{eff} is a skill factor between 0.75 and 2.0 depending on the experience of the leader.

$$T_{report} = 0.45 * T_{prep}$$

$$T_{delay} = 0.15 * T_{prep} + 0.25 T_{rep}$$

The authors of the method claim that it is 90%−95% accurate, which is quite a bold claim given that the estimate is dependent on the assessment of the complexity of the P&IDs. Personally, I am always much happier with estimates based on the numbers of nodes since the node is the entity on which the study is conducted: we know that we will study the same set of deviations in each node so this gives a better 'feel' for the time required compared to the number and complexity of the P&IDs.

One useful aspect of this detailed method, however, is the emphasis it puts on preparation and reporting time; the formulae provided show the significance of these aspects and appear to be reasonably realistic.

3.7.4 Preparation and reporting time

Including preparation and reporting time (time to develop, write and publish the final report) in the overall time estimate is always important for contracted HAZOP leaders. It also draws attention to the resource requirements for the client organisation, in particular for process engineering support in node selection and description. Less has been written about

Table 3.5 Preparation and reporting times [9].

Size of HAZOP	Preparation time (days)	Reporting time (days)
Small	1−2	3
Medium	3−4	5
Large	5−7	7 +

preparation and reporting time than for study time. However, one US rule of thumb set is shown in Table 3.5 [9].

Small, medium and large are not defined in the reference but my judgment would be that the authors have in mind a small study of duration just a few days duration, a medium study taking several weeks and a large to be longer than a few weeks. Nonetheless, I would consider the preparation times in Table 3.5 to be very optimistic. Node selection, marking up and node description in particular can be time-consuming, especially for relatively complex processes and P&IDs; node preparation can take up to 1−2 hours for each node if a comprehensive description is required (as it should be): to include the start and end points of the node, P&ID references, the description of design intent including equipment tag numbers, process operating conditions and equipment design limits. This aspect of recording is described in more detail in Chapter 4.

The time taken to develop, write and publish the report is highly dependent on how much effort is put into developing the report during the study itself. The times shown in Table 3.5 are realistic if the HAZOP leader is organised and diligent at the end of each day of the study. This is discussed in more detail in Chapter 7.

3.8 Preparation you can start immediately

It can be a great relief to get terms of reference that you are happy with agreed and published and, if the study itself is not due to start for some time, it can be tempting to forget about it for a while and work on all the other important things in your in-tray. This would be a mistake. As I have already mentioned, there is the task of trying to get to meet and to know the members of the team. It may take some time for you to make contact and arrange to meet or speak with each team member. You may also want to set up regular communications with the team to discuss progress, to emphasise your commitment to the study, to help them to develop some ownership and perhaps to enlist the help of individuals in finding and bringing together PSI. This will be very important if you need time from the process engineer to assist in writing the node descriptions.

Although you may not have a role yourself in the collection of PSI, it is a good idea to understand who is doing what and to monitor their progress; it is critical that all the required PSI has been assembled and is accessible before the start of the study. However, two aspects of preparation that you must lead yourself and have direct involvement in: are the preparation of P&IDs and the pre-population of the software in which you will (hopefully) be recording the study.

3.8.1 P&ID management

Ideally, the protocol, or at least the principles, for how you will manage P&IDs throughout the study will already form part of the terms of reference, as previously discussed. It is a good idea to have a written protocol for how the P&IDs will be managed so that you can communicate this effectively to the team. An example of a P&ID management protocol is provided in Appendix 8. The critical principle is that a single master version of each P&ID (large size, preferably A1) is owned by the HAZOP leader, marked up with nodes, annotated if necessary during the meetings (with references to recommendations, errors requiring update and other necessary observations), signed off on completion and scanned for inclusion in the final report. If there are a large number of P&IDs this is a significant task and for an extended study there is a risk of losing or damaging drawings. Then there is the task of scanning; this is obviously best done immediately on completion and sign-off for each P&ID to minimise the risk of loss or damage, but there needs to be a process for having frequent access to a scanner that can handle large drawings.

It is worth emphasising that an A3 folder of the P&IDs (marked up with nodes) should be provided for each team member because, while using a large master drawing is a useful vehicle for focusing discussions and bringing team members physically together, individual team members invariably like to have the ability to sit back from time to time and examine their own set of drawings. If it is likely that team members will change during the study then each folder should be assigned to a role (e.g. operations representative) rather than a named individual, with the holder of the role responsible for keeping the folder in good condition and handing it over to their successor or replacement when required.

3.8.2 Software pre-population

The main task of pre-population is creating the HAZOP nodes and adding the detailed descriptions of each node (this is discussed in more detail in Chapter 4). Writing node descriptions during the HAZOP meetings themselves is time-consuming and extremely boring for the members of the team. If you have to do this then you'll be working with the process engineer and recorder while the rest of the team sits and wonders why you are wasting so much of their valuable time (and thinking much less of you as a leader; not what you want at the start of a study with a new team). As a result, when you finally get around to HAZOP itself, they'll be significantly less motivated than they otherwise would be in a well-prepared study, and your task of generating a positive group dynamic and full involvement will be much more difficult.

If you plan to use proprietary HAZOP software like PHA-Pro or PHA-Works then, in addition to nodes and their descriptions, there is an opportunity to pre-populate further information, such as the scope and objectives of the study, the process description, the deviations that will be used, the risk matrix if you are going to use one (building the organisation's own risk matrix into the software can be time-consuming) and the P&ID list, as well as details such as the dates of the planned sessions and the names of team members. And, of course, if you or the recorder have little or no experience with the software then you will need to plan additional time to familiarise yourselves with it. Like writing node descriptions in the meetings, watching the leader and recorder trying to get to grips with the software at the start of the study is frustrating (for you and the team!), stressful for the recorder and reflects badly on both of you (but especially you).

A final detail in relation to recording is defining the policy you will employ for saving of work during the HAZOP meetings, to reduce the risk of losing work. Having to repeat a day of study on account of losing the previous day's worksheets is damaging in relation to the study schedule but completely demoralising for the team and embarrassing for you and the recorder! It has happened many times, so it is worth agreeing with the recorder how this will be avoided, for example by using auto-save, by regular manual saves, by saving a separate file each day (using 'save as' when creating a file for the following day) and saving work overnight in multiple places (network drive, USB drive, e-mail etc.). Please don't let this happen to you!

> **BOX 3.7 Advanced HAZOP preparation checklist.**
> - Signed terms of reference
> - 'Approved for HAZOP' P&IDs
> - Marked-up HAZOP master P&IDs
> - P&ID folder sets
> - Node descriptions
> - Software pre-population
> - Team member familiarisation
> - Recorder preparation

The main points of this discussion are summarised in the form of a simple checklist for advanced preparations is shown in Box 3.7.

3.9 Preparations nearer the study

3.9.1 General considerations

As the start date for the study gets closer, then other arrangements need to be addressed. Table 3.6 summarises the main considerations.

The HAZOP leader may be tempted to assume that these will be taken care of by the host organisation, but this is a risky assumption. To avoid disruption at the start of the study it is advisable to take an interest in all these aspects and to make sure that all arrangements are in place to your satisfaction; with reference to Table 3.6, you should have the contents of the 'Documentation' and 'HAZOP leader' boxes under control through having started your planning early, but the 'Administration' and 'Facilities' aspects can cause enormous amounts of trouble if not planned diligently.

No meeting room booked or team members not notified, or the inability of team members to park or access the building will all delay the start of the study. Poor room layout or climate control, inadequate audio- —visual or Wi-Fi access or the absence of refreshments will disrupt the first session or day of the study, the very time you are hoping to motivate the team and encourage the beginning of team bonding. In the event of difficulties like this, blaming the host organisation, conference centre or hotel will not help; the team will hold you accountable and you will be off to a bad start in developing a motivated high-performance unit!

Table 3.6 HAZOP final preparation requirements.

Administration	Facilities
Room booking	Room layout
Meeting notices	Climate control
Building access and security	Audio—visual
Parking	Wi-Fi, network, video links
	Refreshments
	Name cards
Documentation	**HAZOP leader**
Process safety information (IT)	Recording software/saving protocol
Master P&IDs	Opening presentation
Team member A3 P&IDs	Draft ground rules
Attendance register	Flip charts — actions, parking lot

I once persuaded an organisation to hold a HAZOP off-site, at a city centre hotel, to get the team away from the distractions of their busy plant. We had a great room with excellent catering, but there was no local parking (I hadn't checked) and team members had to pay local traders to watch over their cars while they were in the meetings. They were not happy, my plan had completely back-fired and within a few days we moved back on-site, having wasted the company's money and damaged my own standing with the team and my client.

The 'Facilities' component of Table 3.6 is vitally important in relation to creating an environment in which you can facilitate effectively to enable the team to work creatively and productively, and the opening presentation is your opportunity to engage and motivate the team from the outset. So we'll cover these aspects in a little more detail before summarising the critical success factors for HAZOP preparation.

3.9.2 The HAZOP environment

Fig. 3.7 shows two very different room environments.

It's obvious that the scene on the left is preferable: team members will be grouped round the table in a way that enables them to work closely together in relatively close proximity yet in a reasonably spacious and well-lit environment; the table provides a focal point for the master P&ID. On the right, the table layout means some team members will be a long way from one another, which makes communication between the participants (too many in this example if the room is full) very difficult,

Figure 3.7 Which HAZOP room would you prefer?

Figure 3.8 Which of these is the best brain food?

leaving some team members remote and prone to distractions. Participants will be forced to refer to their own P&IDs rather than using the master P&ID as a focal point for discussion. Facilitating in this sort of environment is extremely challenging.

Visiting the HAZOP room in advance of the meetings gives you the opportunity to create the right layout and environment, test the facilities such as climate control, audio—visual equipment and lighting, and find out how close the restrooms and fresh outside air are in relation to the meeting room to help in the planning of breaks. A visit of this type, including introducing yourself to facilities management (always helpful in the event of equipment breakdowns and the like) is never wasted.

Refreshments are also important in long meeting sessions, whether they are spread over days or weeks; not only providing enough of them but providing the right type. Fig. 3.8 provides a clue.

Regular intake of highly calorific pastries, biscuits or sandwiches is not conducive to alertness, and for many such snacks are difficult to resist over a long day, although once in a while they will be welcomed as a treat or something a bit different. Fruit is obviously much healthier, although it can get monotonous if it the same day after day. A HAZOP team I worked with once actually insisted on fruit being provided each day (I thought it might be joke at first, but it wasn't). The company being unable to organise this, I ended up visiting a supermarket on the way to the study each day to buy the fruit, desperately trying to vary the offering each day; yet another thing to worry about, although in this case it was worth the effort. Of course, if your team wants something else, try and keep them happy... anything to make your job easier and the study more effective and productive! Pay for it yourself if you need to; the benefit in terms of the happiness of your team will easily outweigh the relatively small cost.

In terms of drinks, try to make sure that water is available throughout the meetings and preferably hot drinks too, since you may wish to take convenient points like the end of a node as a point for a break rather than set break times, which may interrupt the flow of the meeting. Participants often like to 'stretch their legs' and get a drink while the meeting is in progress; having to leave the room for this purpose is likely to be disruptive.

Lunch formats are often dictated by the type of meeting room you are using. Rather than having lunch brought into the meeting, a change of scene (hotel restaurant, site canteen) and a walk back in the open air may bring the team back more refreshed.

3.9.3 The opening meeting

The opening meeting is your opportunity to motivate the HAZOP team and launch the study in the best possible way, so it is worth spending time in preparation for it. It is natural to want to launch into the first node of a study as soon as you can, but the time spent in a well-planned and executed opening meeting will lead to a smoother start once you get going.

Box 3.8 shows an example of a typical opening meeting agenda.

Personal introductions for those present is the obvious place to start, but in this agenda they can be kept brief because at Item 9 you will give team members their opportunity to talk in a little more depth, which we'll come back to below. For an introduction to the HAZOP study there is an opportunity for you to ask the sponsor of the study to attend; this is a golden opportunity for a senior member of staff to impress on the

> **BOX 3.8 An opening meeting agenda.**
> 1. Introductions
> 2. Objectives and scope of study
> 3. Project description and design status (if required)
> 4. Process description
> 5. Approach and methodology
> 6. Introduction to HAZOP (if required)
> 7. Recording style and policy
> 8. Process safety information review
> 9. Team member introductions, interests and/or concerns
> 10. Team member responsibilities
> 11. Development of ground rules

HAZOP team the importance of the study to them personally, and thereby provide the expectation that senior management is committed to the study and awaiting its findings with interest. You could also explain yourself how you are going to keep senior management informed of progress (and any potential issues) as the study progresses, and you could invite them (in the presence of the team of course) to attend once in a while for an update, perhaps at the end of each week. The team will appreciate the attention of their senior managers.

For the project, design status and process description it is normally a good idea that this is presented by the project engineer (if present) and/or the process engineering representative, but make sure you communicate your expectations well in advance to enable them to prepare.

You will notice that this style of agenda essentially reviews the important aspects set out in the terms of reference; not the most interesting format perhaps, but it is important that you show your enthusiasm and commitment to the study in the way you discuss them, so prepare a presentation and aim to put your own personal touch to it by emphasising some ways in which you intend to facilitate the meetings (in the way a referee might inform the teams what they expect and how they intend to referee a sports game).

Item 9 in the agenda in Box 3.8 is provided for a special purpose: to give each team member the opportunity to talk on a subject they should be confident in…themselves. HAZOP team members are not always comfortable coming into a meeting room environment, especially if more senior employees are present whom they may not know (or may even fear) and, as we have said, they may have no experience of HAZOP (hence the short HAZOP familiarisation item). You will expect them to speak their mind during the study, so this early chance to speak about themselves and any specific interests and/or concerns they may have is their chance to gain a little confidence in addressing the team; hopefully, this will empower them when they make their first contribution to the technical discussions.

Discussing team member responsibilities is an opportunity to further give team members confidence to engage in the technical discussions. You can give confidence to a process operator by emphasising that you are looking forward to them providing some insight into how the process is, or might be, operated, or a mechanic by emphasising the importance of their insight into the condition of process equipment in a study of an operating plant.

The final item on the agenda in Box 3.8 − the development of ground rules − is to discuss and agree the way in which you are going to behave as individuals and as a team during the study. It is a vital component of the development of the dynamics of the group and we'll discuss it in more detail under the responsibilities of the facilitator in Chapter 5.

The considerations shown in Table 3.6 are presented in the form of a more detailed checklist in Appendix 9, which also includes stationery and

Figure 3.9 A HAZOP facilitation kit.

facilitation aids, some of which are shown in the form of a 'HAZOP Facilitation Kit' in Fig. 3.9.

In organisational terms the HAZOP leader has a lot to think about. Appearing fully in control and well-equipped will give the team confidence; it may even impress them.

3.10 Summary: critical success factors

We've covered a lot of ground in Chapter 3 to emphasise the importance of thorough preparation, to give an idea of the size of the task and the variety of different aspects that need to be addressed: from negotiations with the client over terms of reference to the technical task of detailed node selection and description to the detailed administrative concerns that the meeting room is booked and the coffee is going to be there when the team arrives. We've stressed the need to start preparations as early as you can and to take a personal interest in every detail, however seemingly trivial. By all means ask that it be arranged by somebody else, but make sure that you are personally satisfied that they will deliver on time and in full. Box 3.7 and Table 3.6 provide a summary of the important

> **BOX 3.9 HAZOP preparation critical success factors.**
> - Make sure that HAZOP is the right tool to be using
> - Satisfy yourself that you'll be able to meet the client's expectations
> - Get the P&IDS as soon as you can and define the scope
> - Make an estimate of how much time you need based on your node selection
> - Make sure the team is of the right size and composition
> - Identify the process safety information you need and a process to assemble it
> - Work with the process engineer on node descriptions and pre-populate the recording software
> - Get to know your team and especially the recorder in a mentoring/coaching relationship
> - Always secure terms of reference containing the above and then start preparing immediately
> - Pay attention to the administrative details to make sure the study will start without a hitch

requirements for initial preparation and preparation nearer the study respectively. Appendix 9 can be used as a detailed checklist as the event approaches.

We'll finish Chapter 3 with a summary of 10 critical success factors for preparation — shown in Box 3.9 — that will help you ensure that the study starts without a hitch and progresses effectively from the outset.

Now it's time for the study itself! This is the subject of Chapters 4, 5 and 6. In Chapter 4 we focus on the role of the HAZOP leader in making sure that the study is conducted at least according to recognised good practice, and preferably best practice; this is the technical application side of things. In Chapter 5 we will focus on the subject of facilitation: group dynamics, team development and how you can get the most out of your team, and Chapter 6 will discuss the effectiveness of the study and how you can maximise this.

References

[1] International Standard IEC 61882:2016, Hazard and Operability Studies (HAZOP Studies) — Application Guide, British Standards Institute, 2016.
[2] R.E. McDermott, R.J. Mikulak, M.B. Beauregard, The Basics of FMEA, second ed., CRC Press, 2009.
[3] F. Crawley, B. Tyler, Hazard Identification Methods, European Process Safety Centre, 2003.
[4] T. Kletz, HAZOP and HAZAN, fourth ed., Institution of Chemical Engineers, Rugby, 1999.
[5] B. Kirwan, L.K. Ainsworth, A Guide to Task Analysis, Taylor & Francis, 1992.
[6] W. Bridges, T. Clark, How to Efficiently Perform the Hazard Evaluation (PHA) Required for Non-Routine Modes of Operation Start-up, Shutdown, On-Line Maintenance, Process Improvement Institute Inc., Knoxville, TN, 2011.
[7] R. Jarvis, A. Goddard, An Analysis of Common Causes of Major Losses in the Onshore Oil, Gas and Petrochemical Industries, Loss Prevention Bulletin 255, Institution of Chemical Engineers, 2017.
[8] F.I. Khan, S.A. Abbasi, Mathematical model for HAZOP study time estimation, J. Loss Prev. Process. Ind. 10 (4) (1997) 249−257.
[9] N. Hyatt, Guidelines for Process Hazards Analysis, Hazard Identification and Risk Analysis, Dyadem Engineering Corporation, CRC Press, 2003.
[10] W. Bridges, R. Tew, Optimizing Qualitative Hazard Evaluations for Maximized Brainstorming (or How to Complete a PHA/HAZOP Meeting in One-Third Time Currently Required while Finding More Scenarios), Process Improvement Institute Inc., Knoxville, TN, 2009.
[11] Jenbul Consultants, HAZOP Leader's Training Course, Jenbul Consultants, Otley, UK, 1986 (author's own private copy).
[12] R. Ellis Knowlton, A Manual of Hazard & Operability Studies: The Creative Identification of Deviations and Disturbances, Chemetics International Company Ltd, 1992, p. 64.

CHAPTER 4

Apply best practice

I made it clear at the start that this book does not seek to explain the basic Hazard & Operability (HAZOP) methodology, since this is adequately covered in other publications [1−3]; it is focused on the effective application of the methodology. In this chapter we are concerned with the technical aspects of its application, in which the HAZOP leader must perform as a technical authority in relation to the methodology: the leader must set the standards for how the study is conducted, guide the team in its application and will have the last word in the case of uncertainty or differences of opinion.

This book aims to promote the role of the HAZOP leader as a people-centred facilitator working with and enabling the team to seek consensus, rather than the more traditional model of a general marching the troops through the study telling them what to do. However, when it comes to technical decisions relating to the application of the methodology, the leader will occasionally have to impose rules rather than persuade; to enforce rather than yield.

This chapter is divided into three parts. In the first part (Sections 4.1 and 4.2) we will look at best practice for the identification, construction and documentation of hazardous scenarios through the vehicle of a set of 'golden rules'. In the second (Section 4.3) we will examine and reference some sources of technical guidance for the identification of hazards that HAZOP leaders can use to help stimulate the creativity of the team. And in the final part (Sections 4.4 and 4.5) we will examine some specific technical situations which can lead to debate and differences of opinion, at which point HAZOP leaders should be prepared to step in and firmly explain how the situations are to be handled.

4.1 Constructing scenarios: the 'golden rules'

Unfortunately, the way in which hazardous scenarios are identified, developed and recorded in HAZOP varies widely depending on the standards that are set by the HAZOP leader (or have already been set in the organisation, if the leader is unable to influence these). This has not been helped

> ### BOX 4.1 Golden rules for scenario development.
> - Provide a detailed node description and design intent
> - Look for causes in the node but consequences anywhere
> - Develop ultimate, unmitigated consequences
> - Describe the actions of safeguards
> - Write recommendations in *what-where-why-stand-alone* format
> - Employ full recording

by many of the examples presented in recognised publications [2], some of which we'll refer to. But as HAZOP experience has developed over many years, a broad consensus has emerged as to what could be described as best practice or at least, more modestly, recognised good practice. This is captured in what we describe in this book as a set of 'golden rules' to guide HAZOP leaders in the development of hazardous scenarios; they are summarised in Box 4.1.

Let's discuss each in turn.

4.1.1 Detailed node description and design intent

I stressed the importance of node description and design intent in Chapter 3: to help make sure that sufficient preparation time is devoted to it to achieve the right level of detail in the study. It might seem, on the face of it, to be excessive given that in many HAZOP studies the team may be quite familiar with the process being examined. However, it is fundamental to the effective application of HAZOP methodology because HAZOP analyses deviations from design intent. The design intent is therefore the baseline for the identification and analysis of deviations, and it follows that the team needs to be absolutely clear of what the design intent is: what is a high flow, a low pressure or an incomplete activity? To know that, we have to know what normal is intended to be to a sufficient level of detail.

Table 4.1 is intended to display various aspects of good practice in a format that is often used for describing nodes.

Moving from left to right across the columns, in the 'Node' column it can be useful to give each node a name as well as a number, in order to be able to distinguish between nodes when looking at a long list of them. In the second column, providing P&ID references enables the reader of

Table 4.1 Detailed continuous node description.

Node	P&IDs	Node intention	Design conditions	Operating conditions	Node boundaries
Fuel gas to MX-77 Train 1	071373-3 071373-8 071373-9	Intention: To provide fuel gas to the Gas Turbine Compressor Package Description: Fuel gas is supplied in 2″ stainless steel pipe to the inlet manifold. The fuel gas is used by the Gas Turbine to power the compressor package at the station to compress gas for the pipeline route.	Pressure: 50 barg @ 149°C at TP1 Temperature: 149°C Minimum design temperature: −20°C 2″ stainless steel design pressure: 185 barg	Minimum pressure: 36.1 barg Minimum temperature: 30°C Design operating flowrate: 1.68 kg/s Upstream pressure trip at 44 barg	2″ connection to fuel gas supply at TP1 2″ connection to fuel gas system vent at TP2 2″ connection to fuel gas manifold vent at TP3 M403: 2″ connection to fuel gas manifold drain TP4

the HAZOP worksheets to locate and examine the marked-up nodes on the relevant drawings. In the third column, the node intention is a simple description of what the purpose of this part of the process is and how it is achieved. The example shown is a relatively simple one, but sometimes it may be advantageous to include descriptions of any control loops in the node along with their tag numbers. This sets the scene for the details provided in the design and operating conditions columns: detailing the design conditions enables the HAZOP team to estimate whether or not deviations could lead to loss of containment in the event that these conditions are exceeded; operating conditions (or the operating envelope) indicate(s) the baseline against which deviations are assessed: the start-points for the deviations as it were. The operating conditions in the example shown are fixed points; in some cases it may be appropriate to add to these a normal operating envelope and it also may be appropriate to include some details of the composition of the stream in question. Finally, in the example the node boundaries in the right-hand column provide precise locations for the start and end points of the node: the points on the P&ID between which the node will be identified by highlighting or other means.

For a batch process the node description should be a description of the step in question, describing the objective of the step and how this is achieved. It is good practice to include the status of the process before and after the step, that is, the process conditions (pressure, temperate, composition etc.) as well as the status of valve positions, especially in a programmable logic controlled process where a number of valves may open and/or close to enable the step to be achieved. Table 4.2 shows an example.

In Table 4.2 the information in the starting conditions and ending conditions columns serves two useful functions: it provides additional clarity in relation to what is happening within the step, but it also identifies the equipment and control processes that could possibly go wrong: it provides a useful set of clues. The HAZOP leader can readily apply the batch guidewords to the equipment mentioned in these columns, for example 'NO', 'NOT', 'PART OF' or 'REVERSE' being applied to the opening or closing of valves within the step.

4.1.2 Causes in the node, consequences anywhere

Searching for the causes of deviations is one of the creative parts of the HAZOP process. We want the team to use their imaginations and not be constrained in their thinking but we don't want a free-for-all, so the

Table 4.2 Detailed batch or procedural node description.

Node	P&IDs	Node intention	Starting conditions	Ending conditions	Design conditions
Step 4: De-isolate and start Reactor Circulation Pump 6-P-123	673945 Revision 3.2	To produce a homogeneous mixture of reactants A and B prior to the addition of Reactant C and ensure that the mixture is circulating through the external cooler prior to starting the exothermic reaction.	Reactant inlet valves XCV-1, XCV-2 and XCV-3 closed. Reactor outlet valve XCV-4 closed. Circulation pump suction and discharge valves V1 and V2 closed, cooler isolation valves V3, V4 closed and cooler water supply TCV-1 closed. Pump 6-P-123 off. Flow zero, temperature ambient.	Reactant inlet valves XCV-1, XCV-2 and XCV-3 closed. Reactor outlet valve XCV-4 open. Circulation pump suction and discharge valves V1 and V2 open, cooler isolation valves V3, V4 open, cooler water supply TCV-1 closed. Pump 6-P-123 on. Flow 240 m^3/h, temperature ambient.	Reactor design pressure 10 barg Reactor loop design temperature 80°C

'cause in the node consequence anywhere' rule provides a boundary for our causes: we only search for the causes of deviations within the node that we are studying, but we want to logically track the consequences of deviations to wherever in the process they may manifest themselves. The reason to stay within the node is to maintain a degree of focus and structure; we want the team to examine only the drawing or drawings that relate to the node under study. If team members are asked to think about causes of a deviation within a node that could come from another part of the process it can lead to confusion and loss of focus as more and more P&IDs are consulted and different parts of the process are discussed.

There may well be causes of a deviation in the node under study that come from elsewhere but these should be identified when those other parts of the process are studied. In those other parts of the process, using the 'consequences anywhere' part of the rule should develop the consequence from the cause in that place to wherever it manifests itself. It is quite common that teams find causes of a particular event in various parts of the process. Think, for example, of causes of high pressure in different systems that could result in overpressure of a vent header or causes of overfill in different vessels that could ultimately lead to the overfilling of a knock-out drum. Finding the different causes of events in different parts of the process can present a challenge in Layers of Protection Analysis (LOPA) but the better proprietary recording tools provide the capability to link up causes from different parts of a HAZOP into common LOPA events. But within the HAZOP itself, it is not a problem that causes of particular events are located in different nodes, although it is good practice to cross-check the different scenarios to make sure that the consequences are described consistently.

There is sometimes an exception to the 'causes in the node' part of the rule, and that is at the limits or boundaries of the study. One example would be a raw material being supplied by pipeline from another plant outside of the limits of the study or even a plant in another company. The loss of flow of that material, or a variation in composition, could have a significant impact on the process under examination. So it often makes sense to include loss of raw material supply as a cause of no flow, even though the precise cause of that loss of flow may not be known. If the consequence of loss of flow is significant then a recommendation could be made to investigate the security of supply with a view to improving it if mitigation within the process is not possible. In the same way, a change in raw material composition, or contamination, can have a significant impact on the process and it is important that these deviations are understood. The precise causes — outside of the boundaries of the study — may be difficult to identify but if the consequences are significant then a HAZOP recommendation can be used as a springboard for investigating the issue with a supplier.

A further word about 'consequence anywhere'. This part of the rule also provides a degree of structure to the study. Having identified a cause of a deviation then the team develops the consequence in a logical manner by describing what will happen — step-by-step — in the absence of any mitigation. This is often described usefully as 'telling a story'. Many readers will be familiar with the 2005 Texas City disaster in which an explosion following a release of hydrocarbons from a blow-out stack led to the deaths of 15 workers. A simplified process schematic is shown in Fig. 4.1.

In HAZOP terms, the deviation in question would have been high level in a distillation column node. Working logically from high level in the column in Fig. 4.1 (the Raffinate Splitter Tower), in the absence of mitigation the column overfills into the Reflux Drum, which then overfills into the Blowdown Drum, which eventually overfills and releases

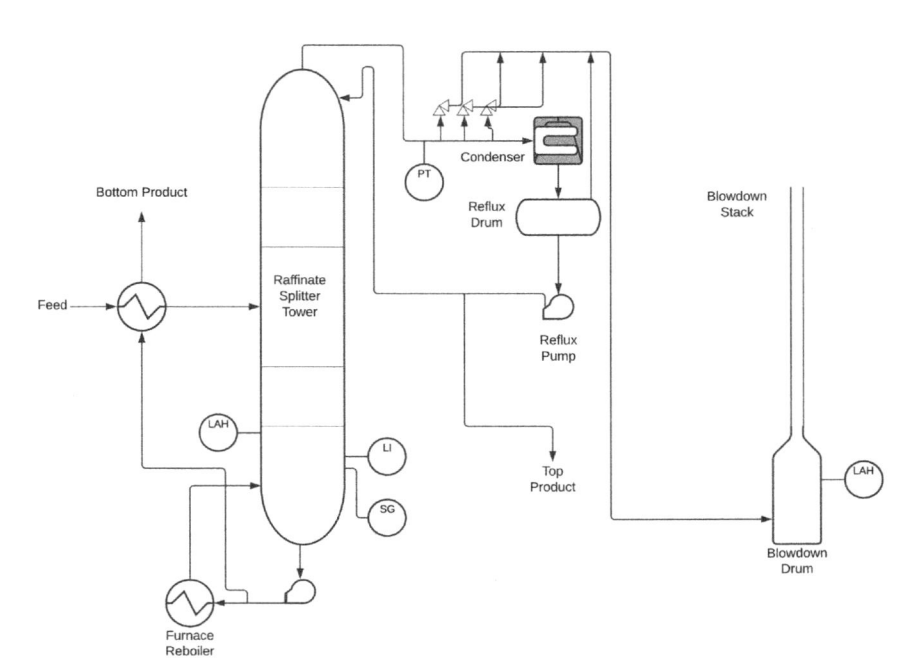

Figure 4.1 Simplified Texas City 2005 schematic. *LAH*, level alarm high; *LI*, level indicator; *PT*, pressure transmitter; *SG*, sight glass.

liquid hydrocarbon into the atmosphere from the Blowdown Stack; this forms a flammable gas cloud, which is then ignited and causes multiple fatalities (in the incident the pressure relief valves were forced open by the hydraulic pressure of the liquid column above them but the final release was the same).

Is this sequence logical? Certainly. Is it credible? Well, it actually happened at Texas City. Would a HAZOP study have developed this consequence? It's easy to say it should do or should have done, but the team would have had to set aside a significant amount of mitigation and the leader would have had to be persistent enough to keep the team telling the story to the very end, despite the possibility of protestations from some team members that the sequence of events is not credible. This takes us on to the other golden rule relating to the development of consequences.

4.1.3 Ultimate, unmitigated consequences

First of all, what do we mean by 'ultimate' and 'unmitigated'? By 'ultimate' we mean the final consequence, which is usually a description of the harm that the event causes. This, of course, could be injuries or fatalities to workers or members of the public, environmental damage, business loss (equipment damage and business interruption, usually expressed financially) and, sometimes, reputational damage such as regulatory action, negative media coverage or public outrage. It is important that we consider all these types of damage and many organisations require this in their HAZOP procedures. By 'unmitigated' we mean the ultimate consequence that will develop in the absence of any safeguards of any kind. This requires the 'suspension of belief' on the part of the team, which is sometimes difficult to achieve in the early stages of a HAZOP with inexperienced team members.

Of course, the events or scenarios that we develop in HAZOP must be credible, and it is important for the leader to be able to keep the team together in the face of members whose imaginations may get the better of them, or others of the opposite persuasion who may look on in disbelief as a scenario unfolds. In the event that the team develops a scenario with an extremely serious consequence but considers the possibility of it happening very remote, it may be useful to develop a second scenario in which the consequence is less severe but the event is more likely. An example might be a toxic gas release that either affects the surrounding population (lower probability, but possible) or doesn't (higher probability). Another example of where different consequences may develop is a flammable gas release that results in a flash fire or an explosion depending on whether it ignites immediately or not. The ultimate consequences of explosion and flash fire may be significantly different in relation to injuries and damage, as may the assessment of mitigation measures, which for an explosion may include the consideration of blast containment.

Describing different consequence/likelihood possibilities is sometimes a good way to manage a disagreement within the team over the credibility of an event, where recording the two different scenarios ensures that all views are represented, but should be used only when the possibility of achieving consensus appears to be remote.

Why is it important that we develop ultimate, unmitigated consequences? One answer is our understanding of risk. If we underestimate the severity of the consequence then we have underestimated the risk, which may have an impact on the way in which that risk is managed. A second answer, which is related to the first, is that we gain a better understanding of the importance of the safeguards we have for mitigating the risk. The greater the severity, the more likely we are to demand further safeguards or more reliable safeguards; the greater the risk reduction required by our safeguards, the more closely we should manage and maintain them.

Developing a consequence by telling the story in a logical, step-by-step way, starting with the cause, is the best way to get to our goal of describing the ultimate consequence of a deviation from design intent. It is sometimes said that HAZOP is not good at identifying major accident scenarios. I suspect that what people mean by this is that the complex nature of many major accidents is not likely to be developed in HAZOP, which is a 'single point failure' methodology considering only one failure at a time. But pursuing ultimate, unmitigated consequences should enable the team to develop such scenarios; the complexity of incident causation often lies in what precipitates the initiating event (the combination of factors that leads to a valve being misaligned, for example) or the failure of the safeguards (the issues that lead to an emergency shutdown valve not being tested or a pressure relief valve being taken out of service), and the underlying organisational failures that allow these things to happen: HAZOP will not identify those because it is not its purpose, but it should identify the events themselves from their proximate causes.

As a leader you will often have to guide the team carefully through the development of the consequence. As stated above, the best way is step-wise and logical. Make sure that the recorder captures the story as it unfolds; this will allow team members to keep up with the story and add their own ideas. The question 'so what?' or 'and then what?' can be powerful here. Level control failure leads to a vessel overfilling. The vessel may overpressure hydraulically if the overflow line is too small (the overflow is a safeguard so should be ignored). Then what would happen? The vessel could rupture leading to a loss of flammable material. So what? Well, it may ignite. So what? People could be present. So what? They might be injured or killed depending on the type of fire. And so we get to our ultimate, unmitigated consequence once we have decided what type of fire might ensue, how many people may be present and whether injuries or fatalities would result. This process does take time, and the recorder must work hard to document it quickly and efficiently, but the result is a comprehensive scenario description that makes it easy for the reader to understand.

Table 4.3 shows a simple example of a consequence of an explosion in a gas turbine that is ultimate and unmitigated.

The HAZOP team might have stopped at the extinguishing of the flame due to the fuel-rich mixture, but the HAZOP leader, with knowledge of the potential for explosions in fired equipment in this type of scenario, has urged the team to consider this possibility. It is the worst case, but it is credible. The team has gone on to consider that workers may be present nearby as the turbine ruptures. The reader should have no problem in understanding what the team was thinking and, if this information is used as part of a resulting recommendation [written in what-where-why-stand-alone (WWWS) format — discussed in Section 4.1.5], then the person who is required to act

Table 4.3 Ultimate, unmitigated consequence.

Deviation	Cause	Consequence
More flow	Flow control valve FCV-0604 fails open	The flow control valve opens quickly leading to a rich fuel gas/air feed to the combustion chamber. The flame is extinguished resulting in a flammable mixture in the exhaust stack which is ignited by a hot surface. Ignition results in explosion and overpressure and rupture leading to 1—2 fatalities in the vicinity of the turbine enclosure and loss of the turbine for several months pending repair or replacement.

on the recommendation will have a clear idea of the scenario and the risk that the recommendation is designed to mitigate.

Should the HAZOP team consider further escalation events, for example the impact from fragments of turbine from the example in Table 4.3? In many cases escalation events are not included in HAZOP as they are often speculative rather than following logically and inevitably from the ultimate, unmitigated consequence. Fragments may or may not escape the turbine enclosure and, if they do, it might be difficult to imagine where they might end up and whether this would cause damage (other than the implicit assumption that they may be the direct cause of the fatalities). In other examples, such as a jet fire that impinges on an adjacent vessel or pipeline, or a pool fire under a vessel, the escalation events are easier to envisage and therefore would normally be included.

4.1.4 Actions of the safeguards

We have spent a considerable amount of time developing causes and consequences and to complete the high-quality HAZOP scenario we want to identify safeguards in a rigorous way too. This means identifying all ways that the causes of deviations can be prevented from developing to the ultimate consequence, or the ultimate consequence mitigated in some way by active measures such as trip systems or pressure safety valves, passive measures such as bunds or blast walls, or administrative means such as process alarms. Having done that, however, we need to make sure that the safeguards we have identified are actually valid safeguards for the scenario: that they will be effective in returning the process to a safe state.

Have you seen a set of safeguards in a HAZOP worksheet recorded like the ones in Table 4.4?

Can we take comfort in any of these safeguards when no justification at all is provided for how they work? A HAZOP team can easily slip into this sloppy way of recording safeguards. They think they know at the

Table 4.4 How do these safeguards return the process to a safe state?

Safeguards
Operating procedure
High-pressure alarm
High-pressure trip
Fire and gas detection system
Emergency response

time what the safeguards are, but are they correct? One way that we can help to test the validity of safeguards is to describe their actions; to describe the way in which they will bring the process back to a safe state. In doing this, it may become apparent that they are not valid safeguards, or that they are only partial safeguards at best (in which case we can note this in the HAZOP worksheet rather than dismiss them completely and demotivate the team member that raised them).

A common example is the claim that a procedure is a safeguard. The first challenge here is to make sure that the procedure in question is actually independent of the cause. If the cause of the deviation is some kind of human failure that the procedure is intended to help prevent, then is it valid to present that procedure as a safeguard? If the procedure contains some kind of cross-checking of the action that causes the deviation then this may be valid, but otherwise it is not. If not following the procedure is a cause, then the procedure cannot be a claimed as a safeguard. If the procedure is actually designed to provide a safeguard against human failure, then what is the specific part of the procedure? Just quoting, 'procedure' as a safeguard is very unspecific. It is better practice to quote the procedure number and specific step of the procedure that represents the safeguard. For example, if the cause of a deviation is a mistake during de-isolation following maintenance, then if the start-up or operating procedure requires a confirmatory check by a second person then that would constitute a valid safeguard. This example is shown in Table 4.5.

In the case of use of a control loop as a safeguard, the same consideration should be given: is the safeguard independent of the cause? If the cause is loss of level control, then that level controller cannot be a safeguard (LOPA challenges these relationships, but we should aim to prevent such mistakes in HAZOP too).

With claims that process alarms provide safeguards, describing the action of the safeguard is particularly important in demonstrating that the safeguard is valid, because doing this can highlight whether the action

Table 4.5 A procedural step as a safeguard.

Cause	Safeguard
Error during de-isolation following maintenance	Operating Procedure ABC-1234 Revision 3 Step 12 requires verification and sign-off of the valve line-up by a second operator using a checklist.

specified as a response to the alarm will successfully restore the process to a safe state in sufficient time and whether the operator has time to respond to the alarm and carry out the specified action. The HAZOP team should not expect to analyse this in a lot of detail (as LOPA would do), but a basic description of the action is useful, as shown in Table 4.6.

In the case of protective devices such as pressure relief valves, their purpose often looks obvious from the P&ID, but the HAZOP team should challenge that they have been designed to protect against the specific scenario in question. This may not always be so, in which case they would not be valid safeguards. An example might be a situation in which the HAZOP team anticipates that a relief situation would involve two-phase flow and yet finds that the pressure relief valve has not been sized for such a case.

For trip systems we should include the manner in which the process is returned to safe state. Table 4.7 contains a description of the action of a high-pressure trip.

By describing what the trip does, the HAZOP team can verify quickly that it is an adequate safeguard.

Moving onto mitigation measures, this is even more dangerous territory, although it is less important if we already have a robust set of prevention and control safeguards. Simply quoting 'fire and gas detection

Table 4.6 Describing the action of an alarm safeguard.

Cause	Safeguard
Failure of temperature controller TICA-456	High-temperature alarm TIA-987 with set point 72°C requires the operator to initiate emergency cooling by opening XCV-654 to prevent the Reactor temperature from reaching 80°C. Response time 20 min available. (Alarm Response Manual Revision 4 Section 3.7).

Table 4.7 The action of a trip.

Cause	Safeguard
Failure of pressure controller PICA-321	High-pressure trip PTZ-888 set point 30 barg closes inlet valves V-32, V-33 and V-34 to the Receiver and vents the Receiver by opening V-35.

system' is not really enough. Is there detection equipment in the place where the gas release or fire would be? What would be the response of such detection equipment? Would automatic sprinklers be activated on the equipment in question or would it simply sound an alarm in a remote control room, in which case could anything be done quickly? In the case of emergency procedures, this simple statement is almost meaningless unless it refers to a specific response plan to the scenario in question, which there may sometimes be, for example in response to a reactor run-away or environmental release. In such cases, it is important to quote the specific procedure and the relevant steps in it to provide justification that it is an appropriate safeguard.

Listing safeguards without testing their validity can lead to compla-cency and underestimation of risk and, as a direct result, failure to gener-ate appropriate recommendations. We want recommendations to be based on a judgement that includes the likelihood of the cause, the sever-ity of the ultimate, unmitigated consequence and the adequacy of safe-guards, and it is important that all three aspects are analysed thoroughly which brings us to the crafting of recommendations.

4.1.5 What-where-why-stand-alone recommendations

Recommendations, especially those requiring additional risk reduction measures to be taken, are the main output from your HAZOP study, and the quality of these recommendations and how they are addressed (if they are!) are critical indicators of the effectiveness of the study. Yet often it feels like getting the study over and done with is more important to the organisation than generating helpful recommendations to make the pro-cess safer. As HAZOP leader it is your responsibility to make sure that recommendations are produced that are useful and written in a way that facilitates them being implemented. The first part of this task is making a good judgement of what the team wants to recommend and the second part is documenting this in a helpful way.

The WWWS rule is a helpful aide-mémoire for constructing a recom-mendation. There is significant work involved in satisfying this standard; it should not be done during the meetings because it will slow them down and bore your team members, so it takes time after study sessions to work on crafting the recommendations in WWWS format. Let's look at three recommendations, all taken from example studies in the literature [1] and then reconstruct them in WWWS format.

Box 4.2 contains a recommendation relating to an offshore oil and gas production pipeline.

Someone with experience in the oil and gas sector would probably guess that the recommendation in Box 4.2 is about the formation of hydrates, and perhaps an experienced process engineer could look at this recommendation and understand what they are being asked to do. But which pipeline and why should I do it? What shall I do with the results of the review? The modelling work that it implies is complex and expensive, so surely it is worth stating why it is being recommended and what should be done with it?

Box 4.3 shows the recommendation in WWWS format based on the information provided in the worksheet containing the recommendation.

It says what is required — the calculation of the temperature-time profile — but it also says what is required depending on the results of the calculation. It tells us where in the process — Block 22 Subsea Pipeline choke FCV-123 — so there is no uncertainty. It tells us why — the risk of

---

BOX 4.2 Which pipeline and why?

Review the temperature/time profile as the pipeline is pressured taking into account the thermal mass of the pipework — the lowest temperature will be at the choke.

BOX 4.3 Subsea pipeline study recommendation in WWWS format.

The Block 22 Subsea Pipeline HAZOP has identified a concern that when the choke (FCV-123) is initially opened the temperature may fall sufficiently to cause the formation of hydrates, which in turn could lead to pipeline blockage and potential overpressure of the line (G1 12″ 15CS), which could lead to loss of containment, fire and multiple fatalities in the Well Bay. It is recommended that the temperature-time profile for the pipeline during choke opening is calculated, taking into account the thermal mass of the pipework. If there is a risk of hydrate formation and blockage, then the adequacy of the methanol injection regime should be assessed and, if necessary, additional safeguards should be developed.

hydrate formation and the potential consequences — so the reader can judge the importance of the recommendation. And it is stand-alone, meaning that it can be read and understood without recourse to the HAZOP worksheet. This last point is so important because of the modern tendency to load actions and recommendations into tracking databases without bringing any context with them. We write the context into the recommendation to make it stand-alone. The second version probably took 15 minutes to write compared to the first version's 1 minute, but it has made it more likely that the recommendation will be understood and acted on.

Box 4.4 contains a recommendation from the HAZOP study of a chemical reaction.

We can perhaps understand what this is about: avoiding dependence between two temperature probes. But where in the process are they, which probes are they, what is the consequence of their dependent failure and what happens if it proves impossible to separate them? Box 4.5 shows this recommendation reconstructed in WWWS format based on the information in the HAZOP worksheet from which it was drawn.

BOX 4.4 Which temperature probes and why separate them?

Check whether it is possible to physically separate the two temperature probes (control and protection) to reduce common cause effects.

BOX 4.5 Temperature probe recommendation in WWWS format.

The XYZ Batch Reaction HAZOP has identified a concern that failure of temperature transmitter TT-31 (transmitter falsely reading low) for the reactor temperature controller TICA-31 could lead to overheating of the reactants and runaway reaction, which could result in overpressure, reactor rupture and potential fatalities to operators as a result of toxic exposure. The high-temperature alarm for the reactor, TIA-32, is provided by transmitter TT-32, which is located in the same pocket (sheath); the two transmitters could therefore be subject to common cause failure. It is recommended that this reaction runaway scenario is subjected to LOPA to ensure that adequate protection is provided.

The location, equipment and instrument numbers are all mentioned to avoid any doubt. The risk is described and the recommendation is expanded to include LOPA, which would help to identity the impact of dependence of the instruments if they could not be separated. The new version is now self–contained or stand–alone.

Box 4.6 shows a final example, from the same source: it is the worst of them. It is not possible to guess the context of this recommendation without going back to the HAZOP worksheet.

It is a rhetorical question, but what is the probability that a recommendation written in this form will be acted upon? What is 13.2? Not only would the person required to address this have to track down the HAZOP worksheet, but he or she would also have to find the reference 13.2 and work out the connection. Box 4.7 shows an alternative way of expressing this recommendation in WWWS format.

This example took considerably longer to decipher than the other two, demonstrating that the best time to write such recommendations in WWWS format is shortly after the study session in which they were identified!

BOX 4.6 What is this concern with nitrogen?

Review means of displacing nitrogen — a potential contamination in gas as part of 13.2.

BOX 4.7 Nitrogen recommendation in WWWS format.

The Block 22 Subsea Pipeline HAZOP recommends that nitrogen pressurisation be considered as a means to reduce the risk of hydrate formation in pipeline G1 12″ 15CS as the choke FCV-123 is opened at start-up. The HAZOP has identified a potential concern that nitrogen could be subject to adiabatic compression by incoming gas as the choke is opened, leading to very high temperatures and potential damage to the pipeline or its components. In the event that the pipeline is to be pressured with nitrogen (or an alternative gas), it is recommended that the risk of adiabatic compression in the pipeline at start-up is evaluated, and appropriate means of displacing the gas is developed.

One final concept relating to the crafting a recommendation is that of 'SMART actions': specific, measurable, achievable, realistic and time-bound. This concept is widely used in many walks of life, so should we use it in HAZOP? In writing any recommendation, it is always useful to have this concept in mind, but it may not be appropriate to use it. Looking at each SMART aspect in turn, it may not always be possible to be very specific; sometimes a more general recommendation to address a risk is more appropriate so as not to constrain the person who is going to act on it. A recommendation will often require additional risk reduction but that risk reduction — at least in HAZOP — may not be measurable (as it is in LOPA). A recommendation may not be achievable or realistic if the risk is difficult to address which is why in the above examples I considered what might be done subsequent to the required action, depending on its outcome. And as a HAZOP leader it is unlikely that you will be involved in the setting of timescales for implementing your recommendations, although you will likely want to prioritise your recommendations according to risk.

To conclude our examination of recommendation writing, the recommendation in Box 4.8 appears to satisfy most of the SMART criteria.

It looks very specific and, on the face of it, relatively straightforward to implement. However, it is arguably too specific: the HAZOP team has started to redesign the process! The process engineer who eventually has to address the recommendation may well find an easier, simpler, cheaper way to achieve the desired risk reduction. So unless the detail of the recommendation is obvious to the HAZOP team, it should not strive to prescribe precise

---

BOX 4.8 An excessively specific recommendation.

The ABC Boiler Package HAZOP Team recommends that the following steps need to be taken to reduce the risk of loss of primary containment of fuel gas via nitrogen connection points, which could lead to an explosion in the Boiler House and potential fatalities:

1. Remove the nitrogen purge points P2, P3 and P4 on P&ID 46-D-11716 for fuel gas supply to the burners B1, B2 and B3.
2. Provide a check valve and blind flange on nitrogen purge points P2, P3 and P4.
3. Design for gas detection to detect leakage from the nitrogen purge point P1 downstream of the gas safety shut-off valves ESD1, 2 and 3.

risk reduction actions or process design changes; leave this detail to process engineers working for the facility. At its most concise, bullets 1−3 could be omitted from the recommendation, leaving only the first paragraph which by removing the words 'the following' satisfies the WWWS criteria.

To complete our 'golden rules' you won't be surprised by the final rule, having laboured so far in this section to stress the value of creating thorough descriptions of HAZOP scenarios and recommendations.

4.1.6 Full recording

In the early days of HAZOP − before the computer age − study sessions were normally scheduled for mornings only. Notes were recorded by hand, often in pencil on A3 paper, and the afternoon was spent by the leader and recorder typing up the morning's notes before bringing them to the following morning's session to share with the team. Those were the days! Nowadays it's almost always full-day sessions with tidying up in the evening! Because notes were being recorded by hand, it was very difficult to make them comprehensive; this could be done by using a secretary with shorthand skills but that was outweighed by the disadvantages of not having a technically literate recorder... it was tried many times. The concept of 'recording by exception' developed for this reason. Recording by exception means recording only those scenarios that have significant consequences and sometimes, in a particularly extreme manifestation, recording only those scenarios that end with a recommendation! You can see examples of this in the publication from which the recommendations discussed in Section 4.1.5 were drawn [1].

Today, recording by exception is rare and is discouraged for a number of reasons, the main one being that it does not provide a full record of what the HAZOP team discussed and therefore does not demonstrate that a thorough study was performed. Today's technology − including the availability of proprietary recording software − makes full recording more achievable, which in turn has advantages in relation to making it easier to review and revalidate HAZOP studies, using HAZOP worksheets to inform LOPA studies, using HAZOP worksheets as a reference when making modifications and using HAZOP reports to develop training materials and operating procedures.

So what exactly do we mean by full recording? It includes a full description of each node, as discussed in Section 4.1.1, together with a complete and thorough record of each scenario that is developed: all causes, a full description of the consequence 'story', the safeguards and their actions and any recommendations in WWWS format. The examples provided in Sections 4.1.1–4.1.4 show the level of detail that we aspire to, exemplified by the use of tag numbers for equipment and control loops, alarm set points and response times. But it goes further than that. If a scenario is not developed, for example because a cause of a particular deviation cannot be identified, then that deviation is still included and 'no cause identified' or words to that effect are recorded in the 'cause' column. If one or more causes are identified but consequences turn out to be insignificant, then those causes are recorded and 'no significant hazardous consequence identified' is recorded in the 'consequence' column, or words to that effect. Where a scenario is developed but no recommendation is identified, then sometimes 'no recommendation' may be recorded or 'no additional safeguards required'. All this might look pedantic, but it is methodical and provides evidence that the study has been thorough.

Another aspect of full recording is the avoidance of cross-referencing. We saw an example of this in the recommendations in Box 4.6: 'Review means of displacing nitrogen – a potential contamination in gas as part of 13.2'. This forces the reader to find 13.2, identify what it means and relate it to the statement that cited the reference: a recipe for confusion. More common is one scenario which cross-references another, for example, 'consequences as for no flow'. There is less chance for confusion here, but the reader is forced to look for what was recorded under 'no flow'. With modern software it is easy to copy-paste from one place to another to avoid cross-referencing. A smart recorder will quickly identify the need to do this and do it during the meetings without inconveniencing the team, otherwise cross-references can be removed when the worksheets are reviewed at the end of each session or day.

There are some disadvantages to full recording within the study sessions themselves: a little more time and effort is required and the team may have to be a little patient from time to time. However, these are far outweighed by the improved quality of the final product and its usefulness to the organisation as a document to use, follow-up and consult again in years to come.

4.2 A summary of scenario development

Table 4.8 summarises the important considerations for developing hazardous scenarios.

Table 4.8 Scenario development summary.

Causes	Consequences	Safeguards	Recommendations
What could go wrong? Identify ALL the initiating causes of this.	What is the ULTIMATE, UNMITIGATED consequence? Develop this by describing what happens after each cause, in the absence of any safeguards. Cause leads to...and then...and then... (tell a story).	What are the safeguards or barriers? Preventative AND mitigating. Passive (requiring no action), Active (requiring action) and Administrative (controlled by procedure). Describe how the safeguard(s) makes the process safe.	Are the safeguards sufficient to control the risk? If not, make a recommendation to reduce the risk or investigate/ analyse further in What-where-why- stand-alone format

HAZOP leaders need to have these requirements firmly at the front of their minds in every study they facilitate and make sure that they are satisfied consistently. In this way you will be confident that you are applying best practice.

4.3 Tools for stimulating creativity

In the second part of this chapter we will examine and reference some sources of technical guidance for the identification of hazards that HAZOP leaders can use to help to stimulate the creativity of the team.

It is the HAZOP leader's responsibility to stimulate the team to think creatively about the causes of credible hazardous scenarios, to ensure that scenarios are developed fully and logically and that they are technically correct in so far as the team's knowledge permits. Ideally, the leader needs to play no part in contributing to this activity: the leader opens with a question and the team does the rest. In reality things are unlikely to go that smoothly. Teams may, from time to time, find themselves struggling to generate causes of deviations or follow a logical path in the development of consequences. In these situations the HAZOP leader must help them by asking questions that prompt team members to provide the answers, open up the discussion and move forward.

This section introduces three tools that the HAZOP leader may have at their disposal as aides mémoires to help them to find the right question to put to the team. It is not suggested that the leader uses these tools overtly nor that they be shared with team members (which could cause problems if they distract them), but they can be useful prompts for the leader in the event that the team is struggling to identify causes or develop consequences.

4.3.1 Modes of loss potential causes

The tool provided as Appendix 10 [4] provides lists of possible causes that relate to four 'modes' of loss of containment:

- containment lost via an open end route to atmosphere;
- containment failure under design operating conditions due to imperfections in the equipment;
- containment failure under design operating conditions due to external agencies; and
- containment failure due to deviations in plant conditions beyond the design limits.

For each of the modes of loss a table provides a large number of potential causes, some of which are beyond the scope of HAZOP (manufacturing defect or substandard maintenance for example) and some more appropriate to Hazard Identification (blast effects from a nearby explosion for example). But the majority of potential causes are relevant to HAZOP. Take the deviation 'low pressure'. The table offers a number of potential causes of equipment being subjected to vacuum that it may not be capable of withstanding: connection to equipment under vacuum; the movement or transfer of liquids; cooling gases or vapours; solubility effects such as dissolution of gases in liquids. With the table at their side (or better still in the memory!) the HAZOP leader may be able to stimulate the team to think about low pressure by asking questions such as 'Could low pressure or vacuum be generated by...?' or 'Could a movement of liquid generate vacuum?' if team members appear unable to generate such ideas.

4.3.2 Process flow failure modes

Appendix 11 presents a document that has become known as 'The List' [5]. It is based on 20 years of process hazards analysis experience and has developed into an ongoing open collaboration. It is a list of process failure modes for more than 20 types of equipment commonly found on process plant: pipework, pumps, heat exchangers, vessels and so on. So whereas potential causes discussed in the previous section are categorised by mode of loss, 'The List' is categorised by equipment type and so appears more like a Failure Modes Analysis tool. As an example, if the node under consideration contains a compressor, the HAZOP leader could scan 'The List' during discussions and look for possible scenarios he or she could suggest to the team. The leader might think to themselves: 'Have we considered closed suction of discharge valves yet, or sticking check valves, or seal failure?' If not, they can bring these into the discussion by means of questions put to the team.

4.3.3 Consequence pathways

The consequence pathways tool shown in Appendix 12 [6] is a simple logic diagram that develops possible scenarios following a leak of a hazardous substance: flammable, toxic or both. Like the cause identification tools described in the previous two sections, it is more a prompt for the HAZOP leader than a tool to be shared with the team, although it could be of use in coaching the team in ways to anticipate different types of fire or flammable release scenarios that may result in multiple types of fire, since it derives from basic process safety training. As an example of its use, an unpressurised leak of a flammable material could form a pool which is ignited and leads to a pool fire. If it evaporates before being ignited it could lead to the possibility of both a pool fire and flash fire (the flash igniting the pool). Alternatively, a pressurised leak of gas could lead to a fire ball or jet fire depending on when it is ignited or an explosion if there is a degree of confinement (an enclosed or partially enclosed space), whereas a two-phase or volatile liquid release could lead to a flash fire, a pool fire following 'rain-out' or an explosion of there is a degree of confinement. Similar logic diagrams can be found in the literature for these types of events [4]. This type of tool is probably of best use to the leader in thinking, 'Have we covered all the different possible consequences here?' or 'Has the team followed the right logic to the consequence?'

To conclude the second part of this chapter, the HAZOP team will occasionally experience losses in creativity or logic, at which point the leader has to be ready with ideas to raise the energy level and focus the team by asking stimulating questions. The tools introduced here are examples of sources of ideas for the leader. Whether these tools are used or others, it is a good idea to have at hand some prompts to stimulate ideas for causes and help the development of consequences. To reempha-sise, they are not tools that necessarily need to be shared with the team. Giving team members access to the tools could risk making things over-complicated (if a team member starts to use them exhaustively) or confus-ing (if different team members use them at the same time).

4.4 Some technical challenges

In the final part of this chapter we will examine some specific situations which can lead to debate and differences of opinion where you will be called upon to exercise your authority in relation to the application of the methodology; in Section 4.4 we deal with technical challenges relating to interactions between parts of the process and then in Section 4.5 examples that illustrate the need to define consistent policies for dealing with spe-cific types of equipment.

It is often stressed that the HAZOP leader does not need to be a techni-cal expert, at least insofar as their role as the leader is concerned. What we really mean here is that the HAZOP leader is not the technical expert on the HAZOP team in relation to the process under study. That role is played by the process engineer, sometimes supported by a chemist and at other times by a functional expert such as automation engineer. Any questions relating to technology or process engineering should be fielded by these people.

Some of my most disheartening experiences have involved being criticised for not understanding the process sufficiently well. In most of those cases I did understand the process as well as I needed to but as a facilitator I tried to get the technical information from the team rather than suggest things to them. Sometimes this meant asking, 'Can you explain how this works?'; I would often have a good idea of it myself, but I wanted the team to provide the information as the custodians of the process. Sadly, this has sometimes been misinterpreted by team members as lack of understanding.

The HAZOP leader should, however, be able to think through tech-nical challenges presented by the HAZOP methodology, because they are sure to arise. This section addresses the most common technical challenges — those of double jeopardy, dependence and process dependency.

4.4.1 Double jeopardy

It is common is HAZOP to hear the exclamation, 'But that's double jeo-pardy...we don't consider that!' Double jeopardy is a legal term (expres-sing that a person cannot be tried twice for the same crime); perhaps strangely, it is used in a process safety context to refer to multiple, independent events happening at the same time and combining to result in a consequence that is worse than the events themselves. This can be represented in the form $A > X$, $B > Y$, $X + Y > Z$, where A and B are the independent causal events and outcome Z is worse than X or Y. Most organisations stipulate that double jeopardy events are not considered in HAZOP and that the study should be limited to single point failures. But the consideration of what constitutes double jeopardy is not always straightforward and HAZOP leaders need to be ready to guide the team in deciding whether such events truly exemplify it.

As an example, consider the typical distillation column shown in Fig. 4.2, in which the feed flow and reboiler temperature are controlled automatically.

The flow controller on the feed to the column FIC-04 fails open (event A), causing higher pressure in the column (event X, $A > X$), while

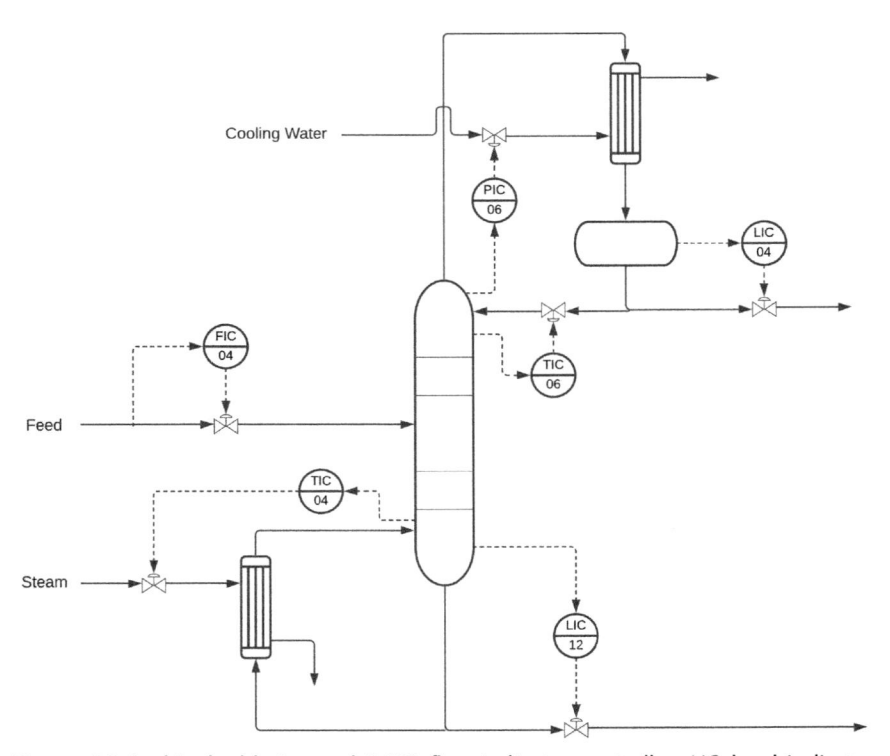

Figure 4.2 Is this double jeopardy? *FIC,* flow indicator controller; *LIC,* level indicator controller; *PIC,* pressure indicator controller; *TIC,* temperature indicator controller.

at the same time cooling to the condenser is lost when PIC–06 fails closed (or the steam control valve TIC–04 fails open; event B), causing the pressure to increase further (event Y, $B > Y$), to the point that the condenser is over-pressurised and fails, leading to loss of containment and potential fire or explosion $(X + Y > Z)$. Is this double jeopardy? If two of the three failures have a common cause, such as a control system failure, an instrument air failure or physical damage because they are in the same location, then this is potentially not a double jeopardy situation and the scenario should be fully developed and documented. If, however, it can be shown that the controllers have different input–output cards in the control system, that the instrument air supplies to the controllers are independent or that the instruments are well separated then double jeopardy could be claimed. In this case it is worth following the full recording policy and pointing out in the worksheet that double jeopardy was considered but rejected and the basis for that decision.

Another possible double jeopardy debate could arise when considering tube failure in a heat exchanger which leads to a toxic process fluid contaminating the plant's cooling water system. A team member may point out that if there is a cooling water leak at that time then plant personal could be exposed to the toxic material (albeit diluted). Other team members may point out that to have a leak at the same time as the tube failure is not credible and represents double jeopardy. However, what if the leak from the cooling water system is already there? A cooling water leak would not be an unusual event on many sites and such leaks are often low on the list of priorities for being repaired. An experienced HAZOP leader will know this (an operations representative might also point it out) and suggest that the team should develop the scenario that the cooling water system is leaking at the time it becomes contaminated; this could be added as an assumption in the study that, in the event that utility systems are contaminated with toxic materials, then they are assumed to be leaking somewhere.

4.4.2 Common mode failure

Common mode failure can be represented by the events $A > X\&Y$: one failure leading to multiple outcomes. As an example, consider an instrument air failure that leads to the closure of a shutdown valve on an oil and gas production train, causing the process to vent to flare. In the process of developing the consequence the HAZOP team identifies that the shutdown valve on a second train is fed by the same air supply. In this case the scenario would develop into the simultaneous relief of both trains to flare. When moving on to consider the capacity of the flare system, it is identified that the flare design is based on single train relief (on the assumption that the shutdown valves on each train had segregated air supplies with accumulators on each valve and check valves in the supplies). So we now have a situation where a single air failure (event A) leads to two trains relieving to flare (events X&Y) which results in the overpressure of the flare system and the potential for a significant release of hydrocarbon and possibility of fire or explosion. This example demonstrates the importance and value of having good process safety information available to the HAZOP team (in this case detailed instrument air distribution P&IDs) and using it to verify what the team thinks would happen.

4.4.3 Process dependency or interaction

Fig. 4.3 shows a crude oil heater for which the heat is supplied from a waste heat recovery unit via a recirculating heating medium.

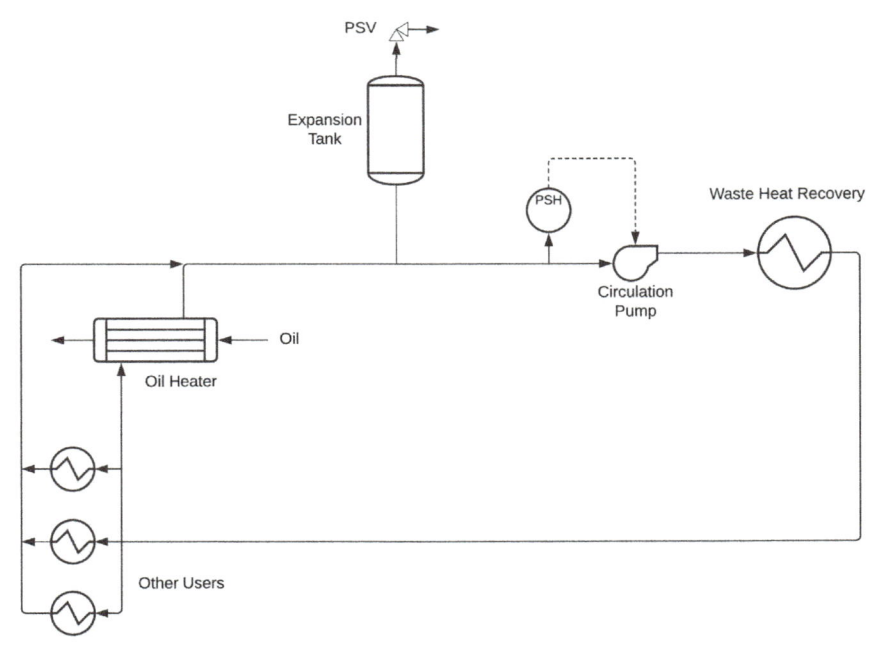

Figure 4.3 Process dependency. *PSV*, pressure safety valve; *PSH*, pressure switch high.

Any reduction in circulation flow of the heating medium, perhaps due to a user coming off line, results in a pressure increase due to the heat gain from the waste heat recovery unit. Such fluctuations are considered to be a part of normal operation, and the expansion tank is protected by a pressure relief valve for that purpose. In the oil heater node of a HAZOP on the system, tube rupture in the oil heater is considered as a cause of high pressure in the heating medium circulation loop. When developing the consequence, the team realises that the high pressure will cause the circulation pump to trip from the pressure switch PSH which in turn causes an additional pressure rise due to the additional heat gain.

The pressure safety valve PSV on the expansion tank is not sized for that case and so we have a vessel rupture scenario. The tube rupture (event A) causes a pressure rise (event X) but also the tripping of the pump (event B) which itself leads to another pressure rise (event Y); the combination of X and Y pressure rises causes rupture of the expansion tank (event Z). The scenario can be represented by $A > X + B$; $B > Y$; $X + Y > Z$.

This example shows the value in working through the sequence of events logically to determine the ultimate consequence, as well as the value good operations knowledge in the team (to identify that the pump will trip) and good process safety information (from which the team can verify the trip setting for the circulation pump and verify the design basis for the expansion tank pressure relief valve).

4.5 Policy challenges

Difficult situations can arise in HAZOP that cause much contentious discussion in meetings but can be avoided with good planning and the development of firm policies to handle them consistently. Such policies, when they are defined, will be stated as key assumptions in the final HAZOP report. Here are some examples.

4.5.1 Overflows

The team is considering the deviation 'high level' in relation to a closed vessel and identifies the failure of a level controller as a credible cause. The consequence is developed as far as the vessel overflowing through an overflow pipe resulting in loss of containment to a bund or dike. This sounds logical, but of course the overflow pipe is actually a safeguard, and so you urge the team to consider what will happen in the absence of that safeguard. At this point, you may have a member of the team mutter, 'That's impossible, the overflow pipe is 6" diameter; that would never block!' So you may have to try and share experiences of blocked overflow pipes from deposits or condensed material from the process, birds' nests, plastic bags tied over the end by painters or slip-plates left in after maintenance. If you get past this, then the next question may be whether the vessel could fail due to hydraulic pressure generated by complete filling or the overfill continuing forwards through the process, as it did in the 2005 Texas City incident discussed in Section 4.1.2. In the case of vessel failure due to hydraulic pressure, you may find that if it is a low-pressure storage tank it may have a weak top seam designed in as a safeguard against rupture, although you would still wish to consider the rupture situation in the event that the safeguard failed. On the other hand, the team may conclude that rupture is not credible because the downstream flow path is of greater area than the incoming flow path. You may wish to record two scenarios: a rupture scenario with remote likelihood and an overflow scenario with a higher likelihood, especially if it helps to prevent or resolve a dispute within the team. It may be possible to develop a policy or assumption that overflows will always be considered to be subject to blocking in order to avoid a dispute the next time a similar situation arises.

4.5.2 Tank bunds

Continuing on the same theme, an overfill or rupture of a tank will lead to a loss of containment for which a bund or dike has been provided. Tank bunds are provided as safeguards against loss of containment of the tank contents at the time of failure; normally to provide 110% of the maximum tank inventory (or 110% of the inventory of the largest tank in a multitank bund) and contain the contents should the tank fail. However, if the flow into the tank is continuous then the bund will provide only a temporary respite against loss of containment on a wider scale (and in some cases could create a further hazard by creating a pool that may be subject to ignition). So overfilling of the bund may be credible and the bund being in a failed state is certainly credible due to cracks in concrete, perished seals at joints between walls and floors, unsealed penetrations cut out for new piping, drain valves (never a good idea) left in the open position in anticipation of a rainstorm, or rabbit holes in earth bunds (another personal experience). So a bund may only offer partial mitigation and may create another hazard if a pool fire is possible. The HAZOP leader needs to ensure that suitable credible scenarios are developed and documented; standardising the approach taken may help to avoid repeated long discussions the next time a similar scenario is encountered.

4.5.3 Positive displacement pumps

Blocking in pipework on the downstream side of a positive displacement pump is a regular source of heated discussion in HAZOP meetings. What will be the ultimate consequence? My preference — if only to simplify matters and avoid disputes — is to assume that the pressure will increase to the point that the system ruptures wherever the weakest point lies; effectively that a positive displacement pump can potentially develop an infinite discharge head. But I have experienced plenty of over-long discussions where views are expressed that this is not realistic because the pump seals will fail before pipework can rupture or the pump has an internal pressure relief valve and so on. If these can be explained convincingly by a mechanical engineer with supporting documentation then fine: reference the information when you describe that safeguard. Otherwise, assume the worst and create a recommendation that the system is provided with pressure relief for the 'dead head' scenario.

The other technical aspect of positive displacement pumps that can generate discussion is reverse flow through the pump. It may be argued that the arrangement of valves within the pump makes reverse flow impossible, in which case record the explanation as a safeguard and provide a reference to the technical justification. Otherwise, it might be argued that even such a valve arrangement may fail, so just assume that reverse flow through positive displacement pumps is credible and treat each case in the same way to save time through the rest of the study.

4.5.4 Nonreturn valves (check valves)

No HAZOP is complete without discussion of nonreturn or check valves, the main concern being their reliability. This is more of concern for LOPA of course, but it is still often a sticking point (!) in HAZOP. Some organisations have corporate policies for HAZOP in relation to the 'credit' given to nonreturn valves, for example ignoring them as safeguards unless there are two dissimilar valves in an arrangement that is subject to routine testing. There is no problem in recording nonreturn valves as safeguards, but the HAZOP team needs to consider them in the context of the scenario: how serious is the consequence that they protecting against and what other safeguards are in place? In other words, is there excessive reliance on a simple check valve against a serious ultimate consequence? In such a case the team may wish to make a recommendation to increase the extent of protection against reverse flow and its potential consequences.

4.5.5 Emergency shut-off valves

A similar type of argument is often made in relation to the integrity of emergency isolation or shut-off valves. Organisations may have HAZOP policies that give no credit for such valves unless they are specified as 'tight shut-off' valves. In a similar way to nonreturn valves the HAZOP team needs to think about the severity of the consequences that they protect against. If the consequences are very serious, then upgrading to 'tight shut-off' emergency isolation valves may be a useful recommendation.

These are just a few examples to illustrate that you will be presented regularly with situations that individual members of the team may look at in different ways, and which may lead to the extended, often fruitless, discussions. You can try and avoid these by looking at P&IDs in advance of the study sessions and trying to anticipate possible issues, such as where you see examples like those described above, or where you see equipment that you may not be familiar with. Within the meetings, always start from the point that consequences are developed in the absence of any safeguards and look to make assumptions or establish policies for dealing with certain situations as a means to simplify the study and avoid repeated unproductive discussions. Remember to record your assumptions in the final report in a key assumptions appendix or in the main body of the report if they are critical.

This chapter has dealt with some technical aspects of applying the HAZOP methodology and, in doing so, we have identified a number of challenges for the HAZOP leader. In the first part of the chapter we mentioned the difficulty in getting team members to identify causes creatively and exhaustively, to develop consequences logically to their ultimate consequence and to 'suspend belief' in the presence of safeguards while doing this. We have also seen the benefits of full recording but suggested that this has the potential to cause a level of frustration among team members who are eager to 'get on with it'. In the latter part of the chapter we have seen technical situations which may generate much debate and differences of opinion. These are all pointers to the fact that to lead a successful HAZOP study the leader has to not only understand the application of best practice but be able to facilitate meetings in a way that team members stay enthusiastic, creative, patient and involved, and collaborate to resolve debates and differences of opinion to ensure that meetings run smoothly and efficiently to produce an effective HAZOP. The importance of facilitation as a skill in its own right justifies its own chapter in this book.

References

[1] F. Crawley, B. Tyler, HAZOP Guide to Best Practice, third ed., Elsevier, 2015.

[2] F. Crawley, B. Tyler, Hazard Identification Methods, European Process Safety Centre, 2003.

[3] T. Kletz, HAZOP and HAZAN, fourth ed., Institution of Chemical Engineers, Rugby, United Kingdom, 1999.

[4] Center for Chemical Process Safety, Guidleines for Chemical Process Quantitative Risk Analysis, second ed., American Institute of Chemical Engineers, 2003.

[5] R.J. MacGregor, Assess hazards with process flow failure modes analysis, Chem. Eng. Prog. 109 (2013) 48−56.

[6] Fundamentals of Process Safety Training, Institution of Chemical Engineers, Fundamentals of Process Safety Training Course Manual, 2016.

CHAPTER 5

Facilitate!

In the first part of this book I emphasised the importance of facilitation to the role of the Hazard & Operability (HAZOP) leader by expanding (in bold text) the definition of HAZOP provided by International Standard IEC 61882:2016 (the plain text). This is shown again in Box 5.1.

The expanded definition is slightly tongue-in-cheek, possibly slightly cynical, but is intended to convey the reality of many HAZOP leaders' experiences: that facilitating groups of people who may ever have met one another and may have different degrees of ownership and interest in the process is an enormous challenge, the success of which is critical in determining the effectiveness of the study.

The expanded definition hints at the history of the HAZOP technique by mentioning that the HAZOP leader may have little facilitation training or experience. HAZOP leaders have often been self-selected as experienced process safety engineers or safety professionals, the qualifications for becoming a leader being experience and technical competence and not necessarily expertise in facilitating groups. And yet in a real HAZOP meeting the leader's process safety experience or technical competence will not always command respect or enable him or her to resolve difficult

---

BOX 5.1 A more realistic definition of HAZOP.

A HAZOP study is a detailed hazard and operability problem identification process, carried out **within difficult time and resource constraints, on a design that may not be fixed and with incomplete information,** by a team, **members of which may have never met one another and have different degrees of ownership and interest in the process, led by a facilitator who may have little facilitation experience, and for a customer whose expectations may not be clear.**

HAZOP deals with the identification of potential deviations from the design intent, examination of their possible causes and assessment of their consequences, **in a creative brainstorming process, the effectiveness of which depends on the manner in which the technique is applied, but which cannot be measured in any scientific way.**

The HAZOP Leader's Handbook
DOI: https://doi.org/10.1016/B978-0-323-91726-1.00012-7

situations: one team member trying to dominate the meeting; team members arguing or being disrespectful to one another; individuals being distracted by their mobile phones; individuals even leaving the meeting and slamming the door behind them. These types of dysfunctional behaviours — and many others — can seriously damage the effectiveness of the study by silencing or alienating individual team members, discouraging participation, disrupting the progress of meetings and wasting valuable time. Resolving issues like this requires facilitation skills: skills that few of us possess naturally and many find difficult to learn other than by the hard way... from bad experiences.

Facilitation forms its own part of this book to make a statement: that the skills of a HAZOP leader should include those of facilitation and that training in facilitation skills should be an integral part of the development of a HAZOP leader. This requires the aspiring or practicing leader to go outside of the technical safety world and into the world of business improvement and change management, where facilitation is valued and widely used. Box 5.2 shows a short definition of a facilitated session from a leading facilitation expert [1].

You can see immediately that this has direct application to HAZOP: meetings are highly structured by virtue of the methodology where nodes and deviations to be used are pre-defined and the role of the leader is to guide the team through these nodes and deviations to arrive at a set of scenarios and recommendations that are understood and agreed by the all team members.

As far as I know it, specific HAZOP facilitation training is not widely available at this time, although some courses do contain a practical element where leaders can facilitate a short session amongst fellow delegates. So the purpose of this chapter is to explain some of the important aspects of facilitation from the wider world in the context of the HAZOP technique and the meeting environment in which it is practiced.

BOX 5.2 Facilitation as defined by an expert [1].

A facilitated session is a highly structured meeting in which the meeting leader (the facilitator) guides the participants through a series of pre-defined steps to arrive at a result that is created, understood and accepted by all participants.

Michael Wilkinson

CEO Leadership Strategies

5.1 The role of facilitator

The International Institute of Facilitation and Change describes the role of a facilitator in three dimensions: *Architect, Pilot* and *Guide* [2]. As *Architect*, the facilitator designs the process which the facilitated exercise will follow and gets to know the group that they will be working with. In HAZOP terms the leader will agree terms of reference with the client or sponsor (scope, styles of HAZOP methodology to be used, team composition etc.) and identify the series of nodes and deviations that will be studied. As discussed in Chapter 3 (Section 3.8) on advanced preparations, getting to know the team in advance, if that is possible, allows you to emphasise the value of their participation and increases your chances of making a swift and productive start to the study as well as maintaining their commitment throughout. In the role of *Pilot* the facilitator prepares the group for the HAZOP meetings, makes sure that the best possible physical environment is provided for the meetings, that all the required materials and information is available and that the team understands and is comfortable with the methodology. This was covered in Chapter 3 (Section 3.9) on preparations nearer the start of the study: preparation of the meeting room; provision of the required process safety information; readiness of the recording hardware and software; the introductory presentation; draft ground rules for the conduct of the meeting. Through the study meetings themselves as *Guide*, the role of the facilitator is to provide a calm and steady presence, giving the team confidence that the study is going well and helping them through difficult situations such as lack of information or time, technical issues such as double jeopardy debates and interpersonal issues such as disagreements.

You can see from this model that, once again, HAZOP fits the more general definition of a facilitated process and therefore facilitation practices from the wider business world should be transferable to it. It is also interesting to note that, while we often talk about facilitation in the context of meetings, the role of the facilitator starts from the very beginning of the project; the planning and preparation will facilitate the execution of the study meetings in the true sense of the origin of the word facilitate...to make something easier (from the Latin *facilis* meaning easy [3]).

5.1.1 Ground rules

Ground rules are an important aspect of successful facilitated meetings so it is worth some explanation. Ground rules are a set of behaviours — agreed in advance by all team members — that are designed to help to regulate the conduct of the meeting and avoid disruptive or dysfunctional behaviour. Some examples of possible ground rules relating to behaviours are shown in Box 5.3.

These and other examples of possible ground rules relating to the application of the methodology are provided in Appendix 13.

The term 'draft' was used in relation to preparation for the meetings because ground rules should not be imposed on a meeting; compliance with ground rules is much likely to be higher if the rules are discussed and agreed at the start by all group members. However, having this discussion from scratch can be difficult and so providing some ideas as a starting point is a good idea; the team may accept them, change them or add new ones; consensus is the key point. This discussion will take up some time at the opening meeting but, if successful, will save significantly more time once the study is underway.

It's worth making a few comments on two of the items in Box 5.3. 'The recorder is a member of the team' originates from several uncomfortable personal experiences in which team members have objected to the recorder participating in discussions, as well as several HAZOP leaders who themselves have not allowed the recorder to participate (even when they appeared to be perfectly able to ask useful questions without it affecting the quality or speed of recording!). 'We will reach consensus wherever possible' aims to restrict the use of the traditional chair's right to have the final say in settling disagreements; exerting this right should be limited to questions that relate to the technical application of the methodology, in the role of technical authority discussed in Chapter 4. At all other times the leader should strive for consensus by trying to resolve the differences of opinion. The subject of consensus is explored further in Section 5.6.

BOX 5.3 Possible ground rules.
- Everyone will contribute.
- The recorder is a full member of the team.
- We will discuss as a team and not split into smaller meetings.
- We will speak one at a time and address the whole meeting.
- We will respect everybody's opinion.
- We will reach consensus wherever possible.
- We will only take calls and respond to e-mails if absolutely necessary.
- We will follow up parking lot issues as fast as possible.

The style of facilitation described here is quite different to the style of the HAZOP leader that I experienced through much of my career, and it is worth underlining that difference in order to emphasise what is now considered to be best practice. In the past HAZOP leaders were selected primarily for their experience and technical competence and often led their HAZOPs in a somewhat 'managerial' way. This often gave the impression that the team was assisting the leader in conducting the study rather than the leader assisting the team as a whole. A model much used in higher education captures this difference as the teacher (or HAZOP leader in this case) as an instructor — 'the sage on the stage' — versus the facilitator — 'the guide on the side' [4].

To illustrate this difference in HAZOP terms, imagine the development of a hazardous consequence from a deviation. The traditional 'instructor' style leader would often develop the scenario based on their experience, particularly if they were familiar with the type of process being studied; the team would observe the leader dictating the logical development of the scenario with individual team members perhaps making occasional adaptations, corrections or challenges. I have looked on as HAZOP leaders have dictated scenarios to their recorders while team members sit and watch without being asked to comment, some of them probably wondering why they bothered to show up for the meeting. In the more modern 'facilitator' style the leader would ask one or more of the team to develop the consequence, perhaps prompting them occasionally with questions such as '. . .and then?', '. . .so what?' or 'have we reached to point where we have described the harm?' The difference is important in relation to engagement and ownership of the study by the team, of course, as well as in harnessing the experience and knowledge of the team to develop high-quality scenarios. The leader must have the confidence that the team has all the required knowledge and experience: trust the team!

This prompts some further consideration of the extent to which the HAZOP leader participates in the discussions themselves rather than simply acting as a facilitator. In the pure facilitator role, the leader will, as we have seen, act as architect in designing the process and preparations for the HAZOP, as pilot in developing ground rules, providing a vision of why the study is important and enthusing the team, and then as guide initiating discussions, keeping them focused, asking challenging questions when necessary and ensuring that everyone remains engaged. With all this to do has the leader the time to actively engage in the discussion, offering his or her own views, giving opinions and occasionally expressing disagreement? Should the leader even try to do this? In my view, certainly: if the study is progressing well, everyone is involved and the group has a positive dynamic then the leader will appear much more 'one of the team' if they can join in with some of the technical discussions. However, there is a danger that the leader becomes drawn into detailed debates, which can easily happen when most of us are interested at a technical level in our studies, and forget the role of facilitator in keeping discussions to the point and moving forward at a comfortable pace. The facilitator must maintain awareness at all times of the pace and quality of the study and the degree of engagement of all team members.

To summarise, the modern HAZOP leader focuses on his or her role as a facilitator, acting not only as a taskmaster (the traditional role, sometimes necessary in certain circumstances) but also as a motivator, bridge-builder, clairvoyant, peace-maker and praiser; all skills that require training and experience to develop and perfect. An analogy has been used of the conductor of an orchestra, who doesn't play an instrument but helps to produce a great performance from his or her team who afterwards can say 'we did it ourselves'.

5.2 Group dynamics

In order to get the most out of the team it is useful for the HAZOP leader to be able to identify the dynamics that develop when individuals interact with one another and to help the team to manage these dynamics. They have been described [5] using an iceberg metaphor. The visible portion of the iceberg above the water is the *content* of the meetings, that is, the contributions of team members in identifying, developing and analysing hazardous scenarios, and the *method* used by the group: how it is organised to undertake this task. For HAZOP this is straightforward, observable and always organised in the same way: it is the application of the HAZOP methodology by a selected group of individuals from defined disciplines, working as a single group in a creative brainstorming mode for the identification of hazardous scenarios and in an analytical mode for the development of hazardous scenarios. Someone not familiar with HAZOP would probably fairly quickly work out what is going on and how it is being done.

The invisible portion of the iceberg represents the *process*: what is happening while the group is working together. These are often unspoken and unnoticed aspects of the group's behaviour such as the atmosphere, the intensity of the dialogues, the degree of participation and depth of engagement of team members, together with other non-verbal behaviours such as body language that might indicate exclusion, frustration or − dare we say it − boredom. An observer would likely be able to detect some aspects of the process by observing a team in action. The HAZOP leader has to try and maintain consciousness of these issues at all times and try to anticipate and intervene before problems start to affect the progress or quality of the study or the participation and enjoyment of team members.

Appendix 14 contains a more detailed description of different aspects of the process, several of which will be explored in the facilitation skills and techniques discussed later in the chapter. This follows a discussion on team development, which is closely linked with group dynamics.

5.3 Team development

Development of a team's *modus operandi* progresses organically but it may take time and may not be a smooth journey, so the HAZOP leader must try to be conscious of how far and how quickly the process is unfolding to ensure that the study develops a good pace as quickly as possible. This is one of the issues that makes it difficult to predict the time required for short studies of only a few days' planned duration. As we have discussed, the HAZOP leader can influence this process by means of thorough preparation, getting to know team members and their backgrounds before the study, providing a stimulating physical environment (including generous refreshments), delivering a motivational introduction so that team members feel valued and believe that they are about to take part in a worthwhile and enjoyable activity, and helping the team to develop ground rules in an open and inclusive way.

Educational psychologist Bruce Tuckman is still quoted as an authority on the development of small groups based on his 1965 model of group maturity [6], often referred to by its four stages: *Forming − Storming − Norming − Performing*. The behaviour of the HAZOP leader in each of these stages is summarised in Table 5.1 and is presented in fuller form in Appendix 15 in the context of a HAZOP team and the style of the leader.

The *forming* stage involves the team developing a shared understanding of the HAZOP study, its objectives and scope and the style in which you, the facilitator, wish it to be conducted. It involves a careful but

Table 5.1 Facilitation styles and group maturity.

Maturity stage	Facilitator behaviours
Forming	Encourage familiarisation, demonstrate openness, invite sharing of concerns, invite participation, provide lots of support, clearly explain the HAZOP technique and the development of scenarios on the worksheet
Storming	Encourage openness, allow conflicts to surface and resolve them, encourage each team member to have their say, build bridges and exploit the complimentary skills of team members, summarise and review often to cement understanding of the HAZOP process and the required quality of documentation
Norming	Let the team develop their decision-making, encourage team members to speak beyond their own role or specialisation, develop team capacity to compensate for any individual weaknesses
Performing	Ease off guiding the team through the process, allow team members to take the lead if they want to

motivational description of the task that the team has come together to perform and often involves some early discussion of the 'golden rules' and how they are applied. This is particularly the case for team members new to HAZOP or for teams composed of members who may have experienced different styles of leadership or different ways of applying the methodology. In this early stage less experienced team members may be reluctant to contribute; this is typically the case with operations or maintenance representatives who may be unused to working in a meeting room environment with members of management. Remember to allow each team member to talk a little (not too much!) about themselves when you ask them to introduce themselves; this is their first contribution and will likely give them confidence to speak once the technical discussions start. Once in progress, expect some early discussions relating to what is meant by ultimate, unmitigated consequences, some difficulty in temporarily ignoring the presence of (often obvious) safeguards or assuming that they fail (whichever works for your team!) and perhaps some frustration with the full recording process, especially as you are also simultaneously coaching the recorder in this early stages of the study.

You may need a lot of patience at this stage. Imagine that your halfway through the first afternoon and the team is still developing the first scenario in the first node! You may also need some belief that the team will get to a smooth, collaborative way of working. But that's the point of the model: to enable you to identify where you are in the development of your team and to have confidence that they will become a high-performing team with your support and guidance.

The *storming* phase may involve differences of opinion, especially if team members are new to HAZOP or to one another. Some individuals may seek to dominate; others may lack the courage to engage. There could be differences in interpretation of how the plant works, descriptions of consequences, the validity of safeguards, how the risk matrix should be used or whether recommendations should be made. You will need to try and encourage and coach the team towards a common approach that you and they are comfortable with, while being patient in allowing some differences of opinion to work themselves through to agreement or at least consensus. It may well involve asking some team members to speak more and others to speak less and reminding the team of the importance of reaching a consensus.

In the *norming* phase the team are converging on a common approach to applying the HAZOP technique and to working with — and accommodating — one another. In this phase you may need to remind the team of the way they have already handled things ('remember in the last node we treated a similar issue like this. . .'), reinforcing where they show consistency ('great, that's exactly how we handled this yesterday') and watch to make sure that all team members are comfortable with the way the technique is being applied. You may ask different team members to lead on the development of a consequence or placement of the scenario on the risk matrix to let both you and the team see that consistency is being achieved.

You will know that you have reached the *performing* phase because you will be playing a much reduced role in proceedings; you may not be saying much at all but the study is moving forward nicely. Team members will be starting to perform your role from time to time and perhaps one of the team may want to spend a session acting as leader. I once experienced this as a contracted leader and remember worrying that the client might hold back payment of some of my invoice! The reality was that the team was actually enjoying the experience and benefiting all the more from it. On the other hand, you may not actually achieve the final phase if your client or team expects you to be an external authority that guides them through the study in a hands-on way and wants to rely on that. After all, they are paying you a lot of money (I hope).

Whether you actively use this model or not, in the first few hours and days of the study you need to be acutely aware of team dynamics and how the team is developing a constructive way of working, encourage this process and be prepared to intervene if you see unhelpful behaviours obstructing the process or inhibiting individuals.

5.4 Individual communication styles

The way in which you can understand team members and try to adapt their behaviours, and your own, to encourage team development is by observing their style of communication. Table 5.2 summarises a simple model designed to assist in this analysis which is further expanded in Appendix 16 [1]. It is important to appreciate that each style has both positive and negative traits; Table 5.2 shows ways in which the facilitator can prevent the negative characteristics of each style.

Project managers and engineers do what they do best because they exhibit the *drive* style of communication: they want to be involved but want to get on and get the job done. This is why they are not always the best people to have in the HAZOP team and also why they are often not inclined to spend much time in HAZOP. They can tend to dominate discussions, potentially putting off other team members from participating but also, in trying to force a fast pace, can risk leaving other team members behind, especially those less experienced in HAZOP. Process engineers that often have intimate knowledge of the process, to a far greater level of detail than other team members, can also display this style; they

Table 5.2 Individual communication styles.

Style	Good and bad	Facilitator behaviours
Drive	Participating but forceful and impatient	Keep sessions fast-paced and well planned
Influence	Creative but talkative	Give chance to talk but stick to ground rules and ask for end point first
Steadiness	Supportive but 'going with the flow'	Check for agreement, praise
Compliance	Constructive but pedantic and detailed	Remind on required level of detail, make recommendations for more study if necessary

don't have time to get a common understanding of the process and ana-lyse it even more! As HAZOP leader you can anticipate that a team member may display this style and make sure that you prepare in detail and plan the sessions well. Don't create opportunities for frustration. Inform the team of what you expect to achieve in each session and watch carefully that all team members are comfortable with the pace. Summarise regularly; this not only slows the pace slightly but enables you to check that everyone is staying up to speed.

Individuals displaying the *influence* style can be enormously helpful in HAZOP: stimulating discussion, providing creativity and keeping energy levels high. However, they can also talk for too long and not listen to others. Good ground rules relating to the (equal) participation of everyone and listening and respecting all contributions can be helpful here, as can asking them to 'start from the end' (start with the point of the interven-tion) and bringing others into the conversation specifically by name, per-haps with a question ('What do you think, Joe?').

The *steadiness* style can be helpful in that people exhibiting this style are often engaged and supportive of the process, paying attention and lis-tening carefully. However, they may also be reluctant to challenge and, at worst, go along with things they may not agree with or disengage if they are not happy with what is being said. The HAZOP leader can try to combat the negative side of this style by watching the body language for disengagement or discomfort and referring to the person by name in order to invite comment and check whether they agree with what has been said.

The *compliance* style of communication exhibits value for the HAZOP process and desires that it be done properly to the right level of detail; a team member communicating in this style can be a big ally in terms of the correct application of the technique but can tend to excessive detail and, at worst, get bogged down in detail and reluctant to accept intuitive judgements, which are vital to making progress in HAZOP. The team may not be able to anticipate with any certainty that a scenario might result in an explosion but if a team member cannot accept that a judge-ment needs to be made and insists on more information to decide the question then this can seriously slow down and disrupt a study. The only time I ever had to ask a client to remove a team member was in relation to a person like this who was completely uncomfortable with discussing risk without a full set of technical data to back it up; we spent a large part of the first 4 days making very little progress, on account of writing recommendations for more and more technical studies as the means to try and make progress and satisfy this particular individual. His colleagues were becoming exasperated and I had exhausted every effort to change his behaviour, including several one-to-one chats between meetings. The client afterwards told me this had happened a number of times before with this person in HAZOP. Thanks for inviting him to my study, guys!

5.5 Facilitation skills

Having discussed the way in which teams come together and develop a way of working, in this section we'll summarise what are often viewed as the key skills for facilitators and look at several techniques and their application in HAZOP meetings. These are some of the topics that you will learn when you undergo training in facilitation, and I'll try and set them in a HAZOP context.

5.5.1 Facilitation skills for HAZOP leaders

Box 5.4 summarises a set of facilitation skills that are highly relevant to HAZOP leaders.

Although they are stated in no particular order of importance, I would say that the first of these — *listening* — is the most important. As we've discussed, the dynamics of the group are all important and it is your job to monitor and analyse these to determine how you can develop and influence them if you need to. Monitoring and analysing the performance of the group is very difficult when you are talking, so try to limit your interventions to when it is necessary to encourage, challenge, summarise or bring someone to the point. Of course, you'll be expected to start off many of the discussions, but once you've asked that first question and someone is talking, start to listen carefully while observing the body

---

BOX 5.4 Some facilitation skills.

- LISTEN . . .only speak when you need to; monitor dynamics
- SUPPORT . . .you are not the star!
- SUMMARISE . . .at key points for understanding and agreement
- CHALLENGE . . .to improve understanding
- LEAD THE PROCESS . . .you are accountable for quality and time-scale
- CREATE A SAFE ENVIRONMENT. . .to allow everyone to share their views
- TRUST THE TEAM . . .they have the knowledge, not you

language around the table. You can try and bring others into the discussions with perhaps just a nod when you think it's time for someone else to build on or challenge what is being said, although you always need to be ready yourself with a question or comment if others don't come in.

At the end of most HAZOP days I am completely exhausted and I believe a major cause of that is the concentration required to listen carefully while at the same time observing the group dynamics and body language, thinking about the quality and pace of the study and anticipating what's coming up next and what potential issues may emerge.

Listening rather than talking provides you with the space to play a *support* role in encouraging and enabling the full participation of all team members. This is difficult for HAZOP leaders who have grown up in studies that are heavily driven by the leader (sometimes encouraged by clients and HAZOP teams that are happy for the leader to take most of the strain), but it makes for a better and more enjoyable study.

Summarising is a vital role for the HAZOP leader and, fortunately, the structure of the technique provides plenty of opportunity for this. You may need to summarise discussions relating to the causes that have been identified or to summarise different views of the consequences of a scenario before converging on an agreed version. The end of each scenario provides a good opportunity to summarise, either before asking the team whether a recommendation is appropriate or after drafting the recommendation, to summarise the whole scenario, perhaps including some encouraging comments to the effect that the scenario has led to a useful recommendation if this is the case. Summarising allows you to confirm the agreement of the team and also acts as a means for team members who may have lost concentration to catch up. The end of each HAZOP session, and before breaking for coffee or lunch, also provide good opportunities to summarise progress and provide ongoing encouragement to the team for the following session. Keep looking for opportunities to summarise!

It is important to emphasise that the purpose of *challenging* what is being said is to improve the team's understanding of what is being discussed and thereby improve the quality of what is recorded in the worksheets. The purpose is constructive and therefore best done in the form of questioning rather than objecting or contradicting. 'Could you just explain how A leads to B?' will work better than 'There's no way that B will happen!' You can set a good example to your team members in this respect and help to establish a healthy dynamic. Team members are likely to quickly adopt your style.

Having so far stressed the facilitation role as a supporting one, we cannot forget that the HAZOP leader will be held accountable for the quality and timely completion of the study, and there are likely to be times where you have to be a little more forceful and *lead the process*. You may have to push a little to improve the quality of the output, such as in getting a little more detail of ultimate consequences or the actions of safeguards, or in trying to inject a bit more pace into the study. Linking back to what we discussed earlier, you can minimise the need for interventions by careful monitoring of dynamics, quality and pace and use summarising to reinforce quality standards ('I think we described the actions of the safeguards really well in that last scenario') and provide early warning of slipping behind schedule ('We hit a few issues in this session, and I'll be looking to move a bit faster in the next session if we can'). Frequent summarising underlines your role as leader.

A prerequisite for the full involvement of team members is that they feel safe to share their views openly, without being treated aggressively or disparagingly. The term *create a safe environment* may suggest an exaggeration of the risk, but speaking to a group of people can, for some, be a very stressful experience, as can the experience of being in a meeting or conference room. So from the very start it is important that you try to create a welcoming, friendly, relaxed feel to the meetings, giving equal attention and status to each team member and openly solicit and value their contributions. This can be very difficult if you are working in a very hierarchical organisation or culture, where employees may be reluctant to speak in the presence of their seniors. (Politely discouraging the direct involvement of senior managers in HAZOP teams, while encouraging their occasional presence to show interest may be the only way to avoid such situations.) But in most cases smiling, looking like you are enjoying yourself and engaging with everyone on a personal level before the meetings and during breaks will help to relax any team members who may be feeling nervous or intimidated.

Last but by no means least of this skill set is to *trust the team*. In HAZOP terms this means trusting that they have the knowledge and experience to provide all the information to enable a high-quality study to be performed. It's your job as facilitator to create an environment in which they will be motivated and empowered to bring all of this to the table! If you feel that the team does not have the appropriate knowledge and experience then it is probably time to raise this with the sponsor and address the issue.

As we have seen, the word facilitate derives from the Latin *facilis* (easy), and this set of skills embodies the essence of facilitation...making it easy for your team to perform. Remember, it's not all about you, it's about your team.

5.5.2 Some techniques to apply in HAZOP

We'll now turn to some specific things you can do as a skilled facilitator in HAZOP meetings. These are aspects of your leadership style that you can think about and practice in every HAZOP session. They relate to opening and closing, maintaining energy, questioning style and keeping the team focussed.

We have already discussed the importance of preparation for the HAZOP study and, as part of that, preparation for the very start of the study where you will meet the team (hopefully not for the first time!) and introduce the study to them in a way that motivates and energises them. Facilitation experts use the mnemonic device *IEEI* or *Inform Excite Empower Involve* [1] to guide their actions at the start of facilitated sessions. First of all *inform* the team of the purpose of the session, for example reviewing the recommendations made during the previous day, the status of the parking lot and what nodes you hope to complete by the end of the day. Second, try to *excite* them by providing a vision of highlighting what you anticipate will be interesting or challenging during the forthcoming session (e.g. a novel piece of technology in the design) and how they will benefit from it (helping the organisation to understand the process of course but also as a learning experience for them as individuals). Third, *empower* them by stressing the importance of their role in the team (for an example how much you are looking forward to them sharing their extensive operating experience with the team). Finally, *involve* them by asking for their own expectations and any concerns they may have. This is all very obvious for the very beginning of a study, but try thinking of it at the beginning of every day, or even every session. For example, at the start of a day you can do it through the process of summarising the previous day and looking ahead to what is to be covered in today's session. Obviously, you will adapt and tailor what you say for each occasion, but if you can employ this technique you will help to maintain energy levels through the study.

Linked to this concept is the volume of your voice as a tool by which to set and maintain the energy level. This is illustrated in Fig. 5.1 [1] in the form of three voice energy levels.

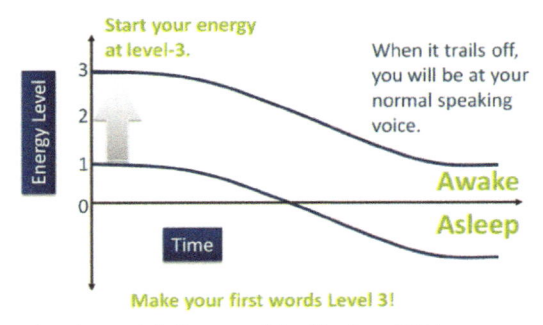

Figure 5.1 Energy levels model. *Secrets of Facilitation, Wilkinson.*

Level 1 is your normal voice as you would employ it in a one-to-one conversation, level 2 is your voice level when speaking to a small room of people around a table and level 3 is your voice level when speaking to a large room of people. During a meeting session your voice level will typically trail off over time, potentially to a point where it drifts into the 'asleep' zone (at level 0) and you start to 'lose' team members (we've all experienced it in meetings). Starting the session at level 1 will risk this happening sooner rather than later. However, starting at level 3 will extend the time at which your voice remains in the 'awake' zone, hopefully to the next planned break. In the same way we applied the IEEI mnemonic down to the level of each session, we can do the same with the voice level model and re-energise our voice at the start of each day, each session, node, scenario and deviation.

The starting technique and voice level will have an obvious impact on energy levels in the meeting. Perhaps less obvious is the way in which you ask your starting questions. We have all heard the HAZOP leader ask 'What are the causes of no flow?', 'What are the consequences of high pressure?' or 'What are the safeguards?' Is this going to simulate creative thinking? The HAZOP leader is asking the team what he or she needs to know, which is logical but far from exciting. Now suppose the HAZOP leader asks 'Imagine how the flow could be stopped?', 'What could happen if the pressure starts to rise?' or 'How would you deal with this?' The idea here is that the question tries to stimulate thinking, perhaps by putting the team more in the situation itself or 'in the moment'. The aim is to fire the imagination a little and get team members more interested and thinking more creatively. The discipline for the HAZOP leader is to put some effort into varying the style in which questions are delivered and, where possible, avoiding simply asking for the information they want

(I have heard many HAZOP team members complain, 'Why does she just keep asking us what she already knows?' or 'Why should I be bothered when he probably knows the answer anyway?').

Having set the discussions flowing with a dynamic first question, we will now be listening carefully to what is said and trying to formulate one or more follow-up questions, should it be required. The style of the follow-up question will depend on what we are trying to achieve. Table 5.3 shows some examples based on six types of questions [1].

To re-emphasise two points made earlier: firstlyeven a challenging question should be phrased constructively and give no hint of disbelief or criticism; secondly, all questions aimed to produce further information should be phrased in a way that seeks to improve the understanding of the team.

We have mentioned summarising as a useful way to cement agreement amongst the team (and allow members a chance to catch up if they've just had a loss of concentration). A related concept is that of maintaining the team's focus by means of *checkpoints* [1]. A checkpoint can be used at any change in the agenda such as a new deviation, a new node or a new plant system. A checkpoint is structured in three parts: *review* what has been done, *preview* what is about to be done and take a *big view* of how this relates to the overall session. As an example:

'We've now covered flow and generated some good scenarios and recommendations' (*review*). 'Let's move on to thinking about pressure;

Table 5.3 Types of follow-up questions [1].

Type	Example
Direct probe to challenge	'Can you explain how will that result in high pressure?'
Playback question to clarify	'It sounds like what you're saying is that the heat exchanger could rupture...is that right?'
Indirect probe to enable team member to clarify	'Is that important because it could lead to overfilling?'
Leading question to seek other ideas	'Are there other factors in your experience of the operation?'
Redirection question to get back on track	'That's a good point. Can we put that on our issues list and get back to the causes of high temperature?'
Prompt question to keep moving	'We've covered the valves on the suction side...what else?'

that may intersect with some of our flow issues' (*preview*). 'After that we'll be well on our way to achieving today's objective of completing this node' (*big view*).

As facilitator you are helping to maintain the flow of the study and create a sense of continuity and progress, rather than just monotonously grinding on from deviation to deviation.

The final technique is that of *closing* and, like the techniques of starting and summarising, it can be applied to a session, a day, a node or a plant system. The discipline is to *review* what has been done, *evaluate* the value of the session, day, node etc. and *remind* the team of the next steps [1]. The idea is to create a sense of achievement as well as anticipation. So at the end of a day's session, you may say something like:

'It's been a long, but I think useful, day. We've made good progress on the Reaction system, had some interesting findings and made some useful recommendations' (*review*). 'I think the quality of our work today has been very good and I think our findings in relation to issue of temperature control will be really useful to the organisation' (*evaluate*). 'Tomorrow we'll aim to complete the Reaction system, which will then enable us to move to the Purification system. Don't forget to attend to your parking lot items, those of you who have them' (*remind*). 'Thanks for your efforts today, have a good evening and I look forward to seeing you tomorrow at 8:45 for coffee and cake before we start at 9:00'.

Much better than, 'That's it for today, see you tomorrow' isn't it? Don't forget to say thank you and finish with some energy and enthusiasm.

5.6 Consensus and disagreement

We have already mentioned that the HAZOP leader should aim for consensus in making decisions in HAZOP. This sounds good, but what is consensus? A lot of people believe that when a group reaches a consensus, then all members of the group believe that the decision is the best decision. If this is right, then it implies that decisions must be discussed and debated until everyone is convinced of the one, best decision, which in turn implies that some discussions and debates will take a long time. In the HAZOP environment, the study needs to progress at a reasonable pace as a sign that the team is working productively; there is no time for lengthy debates about the credibility of a scenario, whether a hazardous scenario might lead to two fatalities or three, or where to position a scenario on the risk matrix.

So we need to take a more pragmatic view of consensus, and the view we should take is that consensus means, for those in the team who do not believe that the proposed decision is the best one, that they will agree to go along with it. From the viewpoint of the facilitator, if a debate looks like becoming protracted then asking a team member if he or she can live with the view of most of the group is far more achievable than aiming to convince them entirely of a position that they initially disagreed with. It's also an effective way to resolve a dispute. It is far easier and more constructive to ask someone if they are prepared to go along with the prevailing view in order to permit the group to move on than to tell them they are in the minority or that you think they are wrong.

From the point of view of such a team member, we want this person to be able to rationalise it along these lines: 'If it were my decision, I wouldn't necessarily go with the credibility of a scenario in which a very unlikely human error results in a major explosion and multiple fatalities. However, I've had the opportunity to share my view and the team have listened to it so, although I have been unable to persuade others of my position, I am willing to accept the team's decision'.

You could also consider generating a recommendation with the objective of shedding more light on the minority view, particularly if the scenario that you are considering has a serious consequence. In this way the majority prevails, but the dissenter can at least feel that their concerns were taken seriously and may actually be warranted should the outcome of the recommended investigation suggest so.

Are there any other alternatives to this consensus-seeking approach? Voting is an obvious one, and is often used in group discussion forums such as committees or councils, but do we really want to resort to voting in a HAZOP meeting? It leaves at least one team member in a 'losing' position and may affect their ongoing commitment to the team and the study as well as the unity of the team; one could imagine an individual seeking revenge after 'losing' a vote, perhaps, by proposing a vote during discussions on a later scenario. Another is leaving the final decision to the HAZOP leader. Again, I would rather have the team broadly agree than be seen myself to be taking sides or taking over. The exception here is where the discussion concerns the technical application of the methodology. In such cases then you are the appropriate authority for such decisions. So, for example, if the team is split over whether to consider the cause of a deviation being a change in composition of a raw material stream from outside of the study boundary, you could legitimately propose that it is appropriate to consider it on the basis that causes can be sought outside of the node at the boundary limits and that changes in raw material composition have been known to result in hazardous situations in the consuming process.

Looking at the wider issue of disagreements, it is inevitable that you will encounter these in every study you attend, and it is easy to imagine that your powers of facilitation will be able to get you through most situations. However, this may not be the case, so it is useful to appreciate that there are different reasons why people might disagree in order to decide if you are able to help to resolve the disagreements. Table 5.4 summarises three levels of disagreement [1].

The simplest disagreements, shown as level 1, relate to a lack of shared information. In this case the disagreeing parties have not heard or understood one another's points and the reasons they have for making them; the disagreement is often based on the parties' assumed understanding of what the other person is saying. Searching for a possible cause of no flow on an existing plant, the process engineer suggests the inadvertent closure of an isolation valve. The operations representative immediately states that such an event is not credible. It looks obvious to the process engineer from the Piping & Instumentation Diagram (P&ID) so he re-states the cause, but the operations representative holds out and you are heading for an impasse. You try to break the impasse by asking a question: you ask the operations representative to explain why he thinks it is not credible to close the isolation valve inadvertently, and he replies that it is located half-way along a 10-m-high pipe rack in a location that is difficult to access and is never used. The process engineer did not appreciate this and therefore did not understand the operations representative's case. Bringing the protagonists to a common level of information resolves the issue. The key

Table 5.4 Types of disagreement [1].

Level	Definition	Facilitator role
1	The people involved lack shared information	Encourage listening by asking questions; slow down to encourage careful listening
2	The people involved have different values or experiences	Understand the issue ('why is this important to you') and create alternatives that combine values
3	Outside factors are affecting the disagreement	Take outside of the meeting for resolution

point here — at this simple level — is to seek the 'unshared' information; in many cases the disagreement will dissolve. (The HAZOP leader would ensure that the recorder notes this in the HAZOP worksheet as a potential cause that is not considered credible on account of the valve's inaccessibility and infrequent use.)

Level 2 disagreements relate to differences in the values or experiences of the proponents. Using the same example, our process engineer has experienced several serious incidents that were caused by operators closing the wrong valve, including errors that were considered very unlikely. At the same time the operator is proud of never having failed to open an isolation valve when required and resents the implication that managers think that operators regularly mis-align isolation valves. Neither is prepared to back down when the location of the valve is made clear. The process engineer believes that if a valve is there, then at some point it will be closed in error; the operations representative that experienced operators don't climb onto pipe racks and close valves without very good reason (although you, as an experienced process safety engineer, know that this has happened!). With this type of disagreement, the HAZOP leader can only try to understand the issues and try and find a way forward that combines the values and experiences of the two parties, in this case by probing a little more about whether the valve is ever operated (at major turnarounds perhaps?), deploying one or two examples where rarely used valves were misaligned at start-ups or perhaps relating some personal experience. It is likely that there is a common value between the process engineer and operator relating to secure isolation valves and perhaps some common ground in considering the case of the valve being misaligned at start-up and then perhaps recommending that the valve be locked open. (The leader might then suggest creating a standing assumption for the rest of the study that inadvertent valve closure will be considered regardless of the location of the valve or its frequency of use.)

It is important to pause here and stress that the HAZOP leader is unlikely to have the time to try and resolve any disagreement beyond level 1; the relatively simple level 2 example just described demonstrates the principle of trying to understand the issue and why it is important to each of the protagonists. Only the best facilitators will succeed in such cases.

For completion, level 3 disagreements are affected by factors outside of the meeting; they appear not to be logical and the proponents do not offer a rationale for their respective positions. Common examples are disagreements based on personalities (the two people dislike one another) or a history of conflict (events within the organisation pre-dating the HAZOP). The only way that these issues can be resolved is by taking them outside of the meeting. Talk to the protagonists at a break and seek agreement to go to a higher level of management to assist with its resolution, especially if there is a risk that there may be further disagreements that could adversely affect the group dynamics in the study.

To summarise, consensus is a worthy objective in HAZOP that can unite the team and avoid disputes. Occasional disputes are probably inevitable though, and the HAZOP leader needs to be able to diagnose quickly if a dispute is going to be able to be addressed within the meeting, or needs to be taken outside.

In the final part of the chapter we'll discuss the subject of dysfunctional behaviour: what behaviours might the HAZOP leader anticipate, how they might manifest themselves and how they can be prevented or mitigated. We'll finish the chapter by considering unexpected 'crisis events' and how we might deal with them.

5.7 Dysfunctional behaviour

As the facilitator we are constantly monitoring the dynamics of the HAZOP team: checking that everyone is participating and that everyone is, if not happy, then at least comfortable with the way that things are going. At the same time we are on the look-out for signs of possible dysfunctional behaviour, that is, behaviour that will disrupt the group dynamic and therefore interfere with the progress of the study. Signs may be expressed verbally, such as raising of the voice or obviously repressed frustration, or through body language, such as sitting back and disengaging with arms folded, raising the eyebrows dismissively in response to something being said or gazing distractedly out of the window. It is important to understand that such behaviour is a substitution for expressing some form of displeasure with the way the meeting is going [1]; it is an attempt to bring about some change by sending a signal to the team, for example that the person disagrees with what is being said, is uncomfortable with the pace of the meeting or is feeling left out. It is also important to understand that dysfunctional behaviour can be conscious or unconscious [1]: the person may or may not be aware of what they are doing. The sudden heavy sigh of frustration is normally noticed by all, whereas the gentle raising of the eyes may be noticed by nobody, including the perpetrator!

Crucially, if we want to be able to prevent and manage it we need to view dysfunctional behaviour as a *symptom* and not as a root cause [1], that is, as an indication of someone's displeasure, not of their personality. Dismissing a team member as 'just an idiot' or 'deliberately awkward' will not help to influence their behaviour.

It is good practice to look around the table on a regular basis and ask yourself whether there are any early signs of dysfunction? Has anyone been quiet for a while? Are there side conversations going on? Is anyone looking uneasy, for example sitting with arms folded or with body leaning away from the group? If this is the case you can plan for a short break to give you the opportunity to talk with the individual concerned before the situation escalates.

5.7.1 HAZOP leader behaviour

Before considering different types of dysfunctional behaviour and how you can deal with it, we need to make sure that we ourselves are not the cause of it! If the first law of engineers is to get through your career without anyone being killed by equipment you have designed or as a result of your action of omission, then the first law of HAZOP leaders should be not to be the cause of dysfunctional behaviour yourself. So how can you yourself cause dysfunctional behaviour? Unfortunately, there are quite a few ways, some of which are summarised in Box 5.5 as 'the seven deadly sins of facilitation' [1].

You will only be able to maintain the engagement of your team members if they feel that their contributions are valued. If a team member suggests a cause for a deviation and this is not recorded alongside those of other members, then this person is less likely to contribute next time and more likely to disengage from the meeting. If 'pump failure' is suggested

---

BOX 5.5 The deadly sins of facilitation [1].

1. Choosing which comments to record
2. Interpreting what is said rather than recording what is said
3. Allowing the team to wander away from the objective
4. Permitting ground rules to be broken without corrective action
5. Losing neutrality and being perceived of favouring member(s)
6. Speaking in an emotionally-charged way or allowing members to do this
7. Allowing an atmosphere of distrust or disrespect to build

by one team member as a cause of no flow, and then another member suggests 'pump stops', don't just dismiss this as a form of failure: consider adjusting the wording to 'pump failure including stopping' to acknowledge both contributions or perhaps go a step further and try and combine each suggestion as 'pump fails due to mechanical or electrical fault' or something similar. It can be tempting to paraphrase what has been said, especially if you (or the recorder) think you can say it more succinctly, but take care to make sure that the contributor can clearly recognise their contribution as it is transcribed; ask the contributor if they are happy with what has been written. This is particularly important if the team member is less articulate or inexperienced at speaking in meetings; such people will be discouraged from speaking if they believe their contributions are being ignored or changed, and this will diminish the quality of the final output.

Allowing an occasional aside, such as a team member relating what is being discussed to an experience they have had, can be useful in adding interest and as a short break from proceedings, but allowing discussions to move away from the subject in hand for more than a few minutes can be frustrating for other team members (an exception might be when you detect the meeting energy level falling and allow a few minutes of humour or anecdote to give everyone a break). You may have a ground rule to try and prevent extended discussions, such as a '5-minute rule' (decide and move on after 5 minutes), in which case don't allow it to be broken, and this goes for all the other ground rules you have established with the team — they were made for a purpose so stick to them.

Favouritism can develop unconsciously if you perceive that the contribution of one or more team members appears to be more useful or succinct than others; the process engineer is invariably the most knowledgeable team member when it comes to how the process works, so it can be easy to look first to that person for contributions. However, this can quickly develop into what looks to other team members like favouritism. Likewise, how you handle disagreements has the potential to be viewed as 'taking sides' if you don't seek agreement and consensus in the ways described in Section 5.6; if you strongly agree with one of the protagonists in a debate it can be difficult to maintain an aura of impartiality.

Given the stressful nature of the role, it is probably no surprise that HAZOP leaders have been known to become exasperated, often where there are time pressures and occasionally with individuals' behaviours. I have certainly raised my voice in an effort to close down debate and move on or increase the pace of the study or expressed exasperation (that deliberately audible heavy sigh) at a discussion that appears to me to be beside the point but won't stop. In both of these cases and many others I have learned the hard way that this is counterproductive. Team members expect the HAZOP leader to demonstrate calmness and control and to use their facilitation skills to overcome problems. As soon as you exhibit emotion your authority is damaged; you look like you've lost control or even that you are looking to blame team members for the way the meeting is going. Any emotional behaviour needs to be stopped immediately, even if that means calling a short break to let matters settle and perhaps discuss the issue with individuals, even it's you that needs a short time-out to calm down and refocus.

The final point in Box 5.5 relates to allowing an atmosphere of distrust or disrespect to build up. You can contribute to distrust or disrespect by exhibiting most of the behaviours discussed above, but there are other ways too. Making mistakes with names (not learning them, forgetting them or repeatedly pronouncing them wrongly) will quickly generate disrespect. Not making allowances for team members' skills in English where it is their second or even third language (such as showing impatience, repeatedly correcting or even mocking) more so. Culturally inappropriate behaviours can be even more damaging, for example failing to take account of religious team members' prayer requirements, which I observed once when a fellow HAZOP leader on an assignment in Malaysia proposed an additional catch-up HAZOP session on a Friday afternoon and then did not understand why this was considered to be disrespectful. And finally, making judgements about the quality of a client's process design, P&ID quality, operating practices or any other aspect of the organisation in which you are working is wholly inappropriate. Your task is to lead a HAZOP study and not to share your own personal opinions of the organisation that is paying you. Sometimes it can be so tempting to express your view, especially in frustration, but don't!

In many cases your team will alert you to any concerns they may have over your behaviour, hopefully quickly and in a discreet way, but this may not always be the case so you may consider asking early in the study if everyone is comfortable with the way you are facilitating.

5.7.2 Role-induced behaviours

Interesting and quite understandable behaviours can be triggered by the role that individual team members play in the HAZOP team. This may not result in dysfunctional behaviour, but it is useful to be prepared for behaviours that may challenge your capabilities as a facilitator and occasionally cause you to have words on a one-to-one basis, especially in the early (forming and storming) stages of the study.

First, let's consider the core roles of process engineer, asset engineer and operations and maintenance representatives. The process engineer is invariably the provider of a large amount of information, explanation and commentary in HAZOP; the study is heavily reliant on this role. Because this individual's level of knowledge of the process is likely to be much higher than other team members', there is a risk that this person tries to progress at a pace that is too fast for the team as a whole or they become frustrated with the pace of the study. You have to be sure that the team is 'keeping up' in the early stages until you have established a pace that everyone is comfortable with. Certain individual process engineers may behave quite defensively in relation to challenges to the design which might manifest itself in defensive statements like, 'There's no way that could happen', 'That's what Operations said they wanted!' or 'I didn't want to design it like that but that would have exceeded the budget'. Asset engineers may likewise be defensive in relation to challenges relating to the condition or maintenance of existing plant or reluctant to embrace recommendation that could require investment or result in more equipment to maintain, which could manifest itself in statements like 'I think the recommendation should be to improve the procedure, not add more equipment'. Operations representatives could take the defensive position of 'That would never happen on our plant', repeatedly urge for additional installed spare equipment, or be reluctant to embrace new risk reduction measures ('If that happened, we'd just sort it out' is common in oil and gas exploration studies in my experience). Maintenance representatives can be similarly defensive in relation to equipment failure ('The equipment could never fail like that') or join the calls for improvements to procedures rather than install more equipment. All such 'role-induced' behaviours need to be handled by impressing the value of the multidisciplinary approach to HAZOP and the importance of seeking a safe process for the benefit of everyone. The alternative is the development of distrust and enmity within the team. Occasional one-to-one discussions outside of the meeting may be required from time to time to point out unhelpful behaviours and seek to address them.

Of the non-core team roles, the presence of vendor representatives is often challenging because they are unlikely to want to make any changes to their equipment. You are likely to hear statements like, 'The equipment could never fail in that way' or 'That's the standard design for this sort of equipment; we have hundreds of these installed all over the world and none of our other customers have ever recommended this change'. Project engineers are likely to be heavily influenced by cost and schedule and this may manifest itself in statements like 'Let's move on...we haven't got time to discuss this; the equipment order needs to be placed by the end of this month'. And finally, senior managers often like to have the last word and may join the end of each discussion with, 'I just want to add that I thought that would be the case'. As we have discussed previously, judicious planning can keep the project engineers' and senior managers' involvement in HAZOP meetings to a minimum.

A final word on the recorder, who also has the capability to create dysfunction by choosing whose comments to transcribe, writing down what they think has been said or should be said rather than what was said, expressing frustration with team members' linguistic abilities and making cultural errors. The HAZOP leader should coach the recorder in this issue before the study but needs to carefully monitor how the recorder is interacting with other team members in the early part of the study and intervene by coaching if necessary.

5.7.3 Dealing with dysfunctional behaviour

In dealing with dysfunctional behaviour we must remember that it is a symptom and not a root cause; it is the behaviour that is dysfunctional, not the person [1]. The person themselves may not be conscious of their behaviour, but they are expressing displeasure, in effect saying, 'I don't like what's going on, but I'm not ready to tell you yet'. If we can think of such behaviour in these terms, then we have a chance of being able to manage it.

Dysfunctional behaviour can take many forms: arriving at meetings late or leaving early; dropping out of the discussion for periods of time; dominating the discussion; telling long-winded stories or moving off the topic in hand; constantly interrupting other speakers; making negative responses to others' ideas or attacking them verbally (or even threatening them physically); whispering to neighbours or holding side conversations; doing other work or playing on a mobile phone or laptop; leaving the meeting in apparent disgust. All of these have been experienced by HAZOP leaders, sometimes several in the same meeting! They are every

HAZOP leader's greatest fear because they threaten the progress and quality of the study and present an enormous challenge to the leader's skill as a facilitator; because they can emerge suddenly with very little notice, the HAZOP leader has to maintain 100% concentration at all times to watch for signs of what might be coming; this is often the main reason why you are exhausted at the end of each HAZOP day!

The management of dysfunctional behaviour, in common with many other unwanted occurrences (including process safety events), has three components [1]:

1. conscious prevention
2. early detection
3. clean resolution

Conscious prevention involves trying to anticipate potential problems during the preparation phase of the study. Fully understanding the background to the study and the reasons why the client or sponsor wants to undertake it can yield important intelligence, for example, are there known concerns relating to the design or progress of the project that could create tensions in the study? Has the study been commissioned on an existing asset because of operational issues or a series of incidents that could become contentious when discussed? A sponsor's desire to have a large number of participants, for example process and discipline engineers from different organisations (client, contractor, vendors for example) could indicate potential 'political' issues that may manifest themselves in the meetings (over and above the inherent difficulty in facilitating a large group of people). Getting to know the team and something of the background of the individuals prior to the study could reveal interpersonal rivalries or antipathies. Limiting the number of team members is always an important concern regardless of the potential for specific issues but measures that could help to prevent dysfunctional behaviour include the development of robust ground rules, interacting with particular individuals before the study and even assigning particular seats to individuals if you think this may help (and not irritate team members). Once the study starts, pay attention to the potential problem issues and use breaks to speak with individuals if necessary.

Early detection requires total concentration and the regular dysfunction checks mentioned at the beginning of the section to look for members who are not speaking, who do not look engaged or whose body language indicates unease with the proceedings. Recognising different team members' communication styles (discussed in Section 5.4 with further guidance in Appendix 16) can also help you to adjust the way you deal with them in order that they are as comfortable and engaged as possible.

Once you detect the onset of dysfunctional behaviour, *clean resolution* entails four steps [1]. First, *approach the individual* involved on a one-to-one basis during a break (if necessary suggest a break if this is urgent). In some cases it may be appropriate to address the group, for example reminding the group about the 'one conversation' ground rule, while not putting the offenders 'on the spot' by naming them (you may need to speak to them individually later). Second, and this is the step that does not come naturally to many of us, *empathise with the symptom* that they are displaying. That may mean saying something like, 'It looks as if you're really busy at the moment' to the person who is distracted by other work, or 'I can understand that you feel strongly about this because you have to operate the plant' to the operator that has raised his or her voice in frustration that they are not being taken seriously. The third step is to *address the root cause* which entails asking a question to understand the issue. To the busy team member this might be 'Is there something urgent that you need to do?' and to the operator 'Is this the type of issue you experience regularly on the plant?' The final step is to *get agreement* on how you are going to move forward. Perhaps you can plan another break to enable the busy member to get their urgent task done or allow them to complete the task before returning to the HAZOP session. To the operator it may be something along the lines of 'I'll make sure that your concerns are given time for discussion if you can be patient and stay calm'.

Appendix 17 contains a guide to the examples of dysfunctional behaviours mentioned above, detailing the common causes of the behaviour, how it can be prevented and how it might be addressed. There is a wide variety of possible dysfunctional behaviours, and they can be extremely difficult to deal with. The guide in Appendix 17 provides an introductory framework but is no substitute for real facilitation training, as much facilitation practice as you can amass and as much feedback as you can get from participants in the meetings that you have facilitated.

5.8 Crisis? What crisis?

A HAZOP leader is likely to experience some level of dysfunctional behaviour in every study they undertake: it goes with the territory. Hopefully, it will never get as bad as physical threats or walking out in disgust, but there is always a possibility of a crisis-type of situation: a situation where the study is abruptly brought to a halt.

Such a circumstance, which you may not agree should be called a crisis, could be a situation in which there is an attack on your own integrity as a HAZOP leader. Perhaps you might make a mistake in the way you apply an organisation's HAZOP methodology to which one or more team members takes exception, or you may be criticised or attacked for the way you are facilitating. I was once attacked by a team member (a young engineer) on the first day of a HAZOP who stood up and accused me — quite aggressively — of applying his company's HAZOP methodology incorrectly. Although I felt that the allegation was wrong, I was mortified and immediately struck by the fear that I might be dismissed and have to pay my way back to the UK from 4000 km away, losing all future work with a major international organisation. What I learned that day was that the correct way to address such an issue is first to acknowledge that the person might be right (difficult as that will be for some!). If you have made a mistake then apologise, correct it and thank the complainant. If you think you are right you can take a break and discuss with the individual or the study sponsor or put the issue to the team for resolution, indicating that your main concern is that things are done correctly and that you are happy to change (or in the worst case step aside!) if this will help the team progress. The key point here is not to let your pride or ego take over; you are only the facilitator. Don't fight back, seek the best solution for the team. In that particular experience, I proposed a time-out and spoke one-to-one with my attacker. When I asked him to explain further, he said he didn't really think I was doing it wrong, but he'd just been on a HAZOP training course at the company's HQ in London and wanted to impress his colleagues that he'd been on this important international trip and now knew all about HAZOP!

Perhaps a more clear and obvious expression of crisis is an unexpected event: something that suddenly takes the team by surprise. Examples might be a team member displaying extreme emotion, such as breaking down in tears or shouting uncontrollably; a medical emergency such as suspected heart attack or stroke; an embarrassing or inappropriate comment, such as a personal insult; a comment that could be perceived as racist or xenophobic, or simply culturally inappropriate. I have experienced all of these in one form or another and heard about a number of other examples, so it does happen and we need to be ready for it. The instinctive reaction amongst the team is likely to be horror or panic perhaps followed by embarrassment in the case of an inappropriate comment which introduces 'the elephant in the room'. The first priority is immediate intervention to calm the emotional person, summon medical attention or stop the source of the inappropriate remarks. It then makes some logical sense to suspend the meeting and reconvene at a convenient time. However, facilitation experts advise [1] that, after the immediate response, rather than dismiss the team ask each person in turn to share how they feel about what has just happened, followed by a discussion of what needs to be done to allow the meeting to continue. In this way the 'elephant in the room' is confronted openly and as a team, and agreed actions can be undertaken in the break that follows, the break being of sufficient length to put these in place. This might be a case of allowing an upset person time to calm themselves and collect their thoughts, finding a replacement team member if a medical diagnosis prevents the victim from continuing, allowing a team member to prepare an apology for inappropriate behaviour or even taking steps to replace that person if that is the only way to bring the team back together.

There have been occasions when an unexpected event has a positive impact on a study. Three days into a 5-day HAZOP that was proving very difficult on account of its 'political' nature (an offshore facility with some significant safety concerns), progress was extremely slow due to some difficult group dynamics. As leader I was becoming very concerned that that team had not reached the 'norming' stage of development and we were at risk of failing to complete the study in the allotted time. In the midst of a tense and quiet part of the meeting the client's Process Safety Technical Authority stood up to stretch his legs and unexpectedly broke wind. After some initial embarrassment all around, the team dissolved into laughter, relationships subsequently relaxed, the study suddenly picked up speed and was duly completed on time. Fortuitous, but not a recommended pre-emptive tactic!

In concluding this section on facilitation, developing good facilitation skills can only come about through training and practice. Don't worry that facilitation training is not HAZOP-specific; the skills are readily transferable and the training experience is valuable as a career development activity in any case (and often fun). Back in the HAZOP world, you'll find yourself giving 100% concentration at all times to what is being said verbally and through body language, anticipating potential issues and taking preventative and mitigating actions to build a team ethos and keep the study moving forward productively. A study can blow up with a very little warning, and at the end of each day you'll experience the tiring effects of prolonged concentration and nervous energy.

Finally, we have discussed how to deal with the inevitable dysfunctional behaviour that you will experience, but I want to finish by emphasising that you can also make the effort to reward functional behaviour, not with favouritism to the individuals concerned but with simple encouragement and thanks. A simple, 'thanks for your contribution' is unlikely to be badly received if it is genuine and not over-used, and is likely to be very effective early in a study where some team members may be reticent to participate. To such a HAZOP newcomer, confidence can be developed by some supportive words like 'Thanks for your contribution this morning. I know that it's your first HAZOP and you may feel a little over-awed but you obviously have a lot of useful knowledge to add to the team and I can see that it's really appreciated'. If you have experienced some dysfunctional behaviour and successfully resolved it, then you can recognise the efforts of the individual involved to reform their behaviour. In the case of the busy mobile phone addict, 'I can understand that it must have been tough for you to avoid your phone during the sessions today, knowing how busy you are outside the study, but I think it really helped the team, so thanks'. And, as we discussed in Section 5.5.2, you can thank the whole team with something like, 'Thanks everybody for your efforts today, we've been really productive' or 'Thanks for sticking with it today everybody. I know it's been a tough day, but I think we're through the hardest part and we'll come back tomorrow with more energy, so have a relaxing evening'. Bringing exciting new refreshments to the next session or 'pizza Fridays' can also be used as a reward and further encouragement to the team. As the man said, accentuate the positive and try whatever you think will work.

Securing the active participation of all your team members, developing a creative and healthy group dynamic, maintaining energy and enthusiasm throughout the study and dealing decisively with nascent dysfunctional behaviour will all have a big contribution to the effectiveness of the HAZOP Study. In Chapter 6 we will discuss the subject of the effectiveness of HAZOP studies in more depth, before going on to discuss the preparation and content of the final HAZOP report in Chapter 7.

References

[1] M. Wilkinson, The Secrets of Facilitation: The SMART Guide to Getting Results with Groups, Jossey-Bass, 2012.
[2] What Do Facilitators Do?, International Institute for Facilitation an Change. https://www.youtube.com/watch?v = UDLGjKBHSXg.

[3] R. Shorrock, D. Butterfield, Latin Dictionary, Penguin, 2007.

[4] A. King, From sage on the stage to guide on the sideVol. 41, No. 1 College Teaching, Taylor & Francis, 1993pp. 30—35.

[5] Fast-track Open Facilitation Skills Workshop Participant Workbook, facilitate this!, 2016.

[6] B.W. Tuckman, Developmental sequence in small groups, Psychol. Bull. 63 (6) (1965) 384—399. Reprinted in Group Facilitation: A Research and Applications Journal? Number 3, Spring 2001.

CHAPTER 6

Focus on effectiveness

Most of us want to look back on the work we've done and say that we've done a good job: that the work has been a success; that we've achieved the objective that we set out to achieve; that we satisfied, or even delighted, our customer. But what do these things mean in relation to a Hazard & Operability (HAZOP) study? We may deliver the requirements laid out in the terms of reference, but how can we answer the question 'Was it a good study?' or 'Was it a successful study?' In Chapter 6 we'll discuss what we are going to call the effectiveness of a HAZOP study. We'll start with what we mean by the term 'effectiveness' and whether we are able to measure it in an objective way. We'll then go on to examine how we can help to maintain effectiveness throughout the study by examining some problems that are common in longer studies relating to attendance, energy and time pressure, and then the circumstances that might lead us to terminate a study prematurely in order to prevent us from delivering something that fails to meet the terms of reference or compromises our own professional standards. In the final part of this chapter we'll examine some ways in which we can assess the effectiveness of a study so that we say afterwards, 'Yes, we have delivered a successful study and this is why...'

6.1 What do we mean by effectiveness?

A 1996 UK Health & Safety Executive report on quality assurance (QA) in relation to HAZOP [1] states, 'A HAZOP can be said to have been effective if the two goals of identifying realisable hazards and executing desirable actions are achieved'. Looking at the second part of the statement first, the subject of desirable actions was discussed in Chapter 4 in relation to the quality of HAZOP recommendations; its inclusion in this statement is useful because it emphasises the point that the HAZOP study is only really complete when the recommendations to implement further risk reduction have been addressed fully. A thorough HAZOP study is of little use if its recommendations are not acted on; we can maximise the

The HAZOP Leader's Handbook
DOI: https://doi.org/10.1016/B978-0-323-91726-1.00013-9

chance that they are acted on by crafting high-quality recommendations in what-where-why-stand-alone format.

The other part of the effectiveness statement quoted above is the first goal: how can we say that we have achieved the goal of identifying realisable hazards? We can count the number of hazards that we have identified but how do we know how many hazards we should have been able to identify? This is the essence of effectiveness: it is the extent to which we identified all of the hazards that are actually present. Of course, the honest answer is that we don't know ('we don't know what we don't know' could be considered as a motto for HAZOP) and therefore we are not able to state in an objective way how effective our study has been.

The 1996 UK Health & Safety Executive report [1], however, makes a positive assertion in relation to HAZOP's inherent effectiveness:

> One attractive feature of HAZOP is the high level of completeness of the deviations considered. This leads, potentially, to a high level of completeness of hazards identified. Whether this completeness is achieved in practice depends critically on the skill of the leader and team in interpreting the guide words.

The premise here, which is the entire premise of this book, is that if a skilled leader with a competent team diligently applies the methodology then the study will be effective in that it will identify a high proportion of the hazards. Over the years I have heard plenty of experienced HAZOP leaders go further than this and say things like 'A good HAZOP should identify 90% of the process hazards' without offering any objective evidence, and without tackling the issue that the unidentified hazards might be the most significant ones. Such faith in the methodology is regularly voiced following major incidents, such as this quotation from the Commission charged with investigating the 1998 Longford incident in the state of Victoria, Australia [2]:

> It is inconceivable that a HAZOP study of GP1 would not have revealed factors which contributed to the accident which occurred on 25 September 1998.

Inconceivable? It is interesting to note that the 1996 UK Health & Safety Executive report [1] states 'The role of HAZOP in the avoidance of major incidents is limited because the scenarios in which major accidents arise often abrogate the assumptions underlying HAZOP' (maintenance, competence, training etc. are all assumed to be in place and functioning correctly).

So is there any objective evidence to indicate the effectiveness of HAZOP? A 2019 paper on the subject of repeat HAZOP studies [3] presents data from three process units on a UK oil refinery site across four cycles of HAZOP studies (the initial study and three subsequent repeat studies several years apart). It shows that, as HAZOPs are progressively repeated through the life of the facility, the number of risks that are discovered at each repeat study or cycle decreases, which is what we would hope to find, but also that a significant number of hazards are found at each subsequent revalidation or re-HAZOP. Obviously, some of these newly discovered hazards will be the result of modifications or 'creeping change', but it is likely that others will be pre-existing hazards that were not discovered at previous studies. This is a strong justification for repeating or revalidating HAZOP studies throughout the lifetime of a facility, both in terms of combatting creeping change and progressively increasing the effectiveness of the facility's HAZOP. Taking a simplified view of the number of hazards identified at the first HAZOP compared with the total number of hazards identified across four cycles, and assuming that the hazards identified after the first cycle were pre-existing, it can be inferred that up to around 80% of the total hazards are identified at the initial HAZOP study (an understatement when allowing for hazards introduced by creeping change). This gives us no cause for complacency and plenty of motivation to examine ways in which we can find evidence to assure us that a study has been effective (especially since those who commission resource-intensive studies might be less than impressed if they are told that the study might only identify 80% of hazards). We'll explore how we might look for evidence after first discussing how to avoid or overcome the most common threats to effectiveness, especially in relation to more prolonged studies.

6.2 Maintaining effectiveness: avoiding or overcoming common problems

Three problems that can severely impact the effectiveness of a study are poor or changing attendance, reduced energy levels ('burn-out') and changes forced by time pressures to complete the study. These problems can affect short studies as well as longer ones and, while they are more likely in longer studies, if they emerge in shorter studies there is less time to react and respond to them.

6.2.1 Maintaining attendance

Poor or varying attendance will disrupt the team; even one change of personnel between sessions is likely to disrupt the group dynamic and force the team back into the 'norming' phase of team development at best. It has an immediate impact, with you or other team members having to explain where the study is up to at the start of the session when a new team member first arrives or an absent member returns; in the worst case the study can go backwards when the new or returning member looks at the previous session's worksheet on the screen and exclaims, 'Who said that? That's nonsense!' forcing the team to go back and review what you thought was already agreed. If this happens regularly then productivity will be lowered and the more 'loyal' team members are likely to become de-motivated, 'What's the point of us coming along each day when the others just pick and choose or send reluctant deputies?' Moreover, from your own personal point of view this will reflect badly on you as a leader and facilitator and further demoralise team members. It is potentially a downward spiral. . .

We have already discussed the preventative measures relating to attendance in Chapter 3: securing management sponsorship and interest and having named team members (including core team members) and deputies in the agreed terms of reference. In relation to the study as it proceeds, you can invite senior management to attend from time to time and encourage them to acknowledge team members' personal commitment on a regular basis. Of course, you can also do this yourself. You can regularly summarise progress, highlight interesting findings and thank team members for their contributions to help them develop a commitment and loyalty to a project that is making good progress and making a difference. And, moving on to the issue of energy, you can create a lively environment that team members will look forward to and enjoy. If core team members are not attending, or if attendance is changing at each session, you may need to seek an intervention by senior management. Make sure that attendance at each session is recorded and make sure you have the data to demonstrate the issue if you think an intervention is necessary. A percentage attendance for core team members (the process engineer has attended 60% of the sessions) or a measure of the turnover of team members (the team composition is changing about 20% each week) are examples of the way in which you could present objective data.

6.2.2 Maintaining energy

'Burn-out' can be a major threat to longer studies, but can also be a factor in shorter, intense studies that are under time pressure or dealing with difficult issues. Like poor attendance (and often linked to it), 'burn-out' will lead to a downward spiral of productivity, lower quality and delays.

One thing most HAZOP leaders have learned from bitter experience is that lengthening the working day or the number of days of study per week in order to speed up the progress of the study rarely works, and may have the opposite effect. So plan the study with no more than 4 days per week and a maximum of 6 hours of study per day to give team members the opportunity to get sufficient rest and attend to their 'day jobs' outside of the meetings. Stick to this and make sure that sessions start and end on time. If you anticipate the need to extend a daily session slightly to complete a particular part of the study then signal this to the team as early as you can, for example by lunchtime, to give them a chance to change any arrangements they may have. Ask their permission to extend the session, and make sure that the extension is as short as possible. If participants are coming to sessions without being sure when they may end you will severely test their commitment and may put them under considerable stress, for example if they have caring responsibilities outside of work.

Within the sessions, take regular short breaks and remember the energy level model (Fig. 5.1 in Section 5.5.2), re-starting each session at a higher energy level and maintaining consciousness of the group dynamic. Is your team enjoying itself? Fun and HAZOP are not often mentioned in the same breath but there is no harm in thinking about what might increase the level of enjoyment. Occasional 're-charge' activities such as showing video clips and sharing anecdotes from team members (especially if they relate to the scenarios you are studying), or discussing learning from major disasters from history that are relevant to the topics under discussion ('safety moments') can give the team an informative small break and give you an opportunity to raise the energy level again when the study recommences. Taking a short time-out can often increase productivity again afterwards, especially if refreshments and/or fresh air are readily available. Organising competitions such as best word or phrase used during the day or challenging the team to bring an unusual or obscure word into the HAZOP worksheets during the day ahead ('Can we get the word "anatomy" into the worksheets today?') can help to keep team members alert while creating a sense of fun. And, of course, providing a constantly changing variety of refreshments never fails to get a positive response; sucking on a mint or fruit sweet can be a useful way for team members to sharpen their concentration when they sense a loss of focus.

One final energy-sapping factor is lack of consistency in the way the methodology is applied that causes team members to work harder mentally. If a session is feeling a bit bogged down it may be tempting to switch to a node in a different part of the process as a means to create a change of scene or change the order of application of the deviations from one node to the next to create a bit of change. In my experience, more often than not this will be resisted by the team; I have found that teams like to get into a rhythm and know what's coming next. Much better to take a short break or try to create some fun than have your team guessing which way you might lead the next node.

6.2.3 Managing time pressure

It is rare for a HAZOP leader to feel comfortable about the amount of time available to complete the study, and if you do feel comfortable one day, the next day could be very different for many diverse reasons that have already been covered. The best forecasting and scheduling can quickly be de-railed for reasons outside of your control: reduction in the budget allocated to the study; the need to re-deploy team members to a more urgent issue within the organisation; a change in business priorities. Or the pace of the study could be lower than you had expected and planned for. The possibility is always present that you will be put under pressure to complete the study earlier than planned for reasons external to the study or to catch up time if the study is behind schedule.

The best insurance policy to avoid being put under time pressure is to have your time estimate agreed in the terms of reference and then to carefully track progress against the plan from day to day, using the achieved rate of progress (e.g. nodes per day) to project ahead how much more time you need to complete the study. Appendix 18 provides a simple HAZOP tracker tool which measures the rate of progress in nodes per day and uses this number, in conjunction with the number of remaining nodes, to estimate the number of days required for completion. Obviously, this is more useful in longer studies, but it is a good practice after every session or day to estimate how much longer you will need to complete the study. You could be asked this at any time by your client or sponsor, so be prepared: any hesitation in answering might suggest that you are not in full control of events. It is also helpful to the team to know for how much longer their commitment is required. Like the parent driving their kids on a long car journey, expect the question 'When will we get there?' to be asked with increasing frequency as the study progresses!

Appendix 19 provides an example of a HAZOP progress report template. It includes a section on progress to-date and anticipated completion, similar to that described above and shown in Appendix 18, but goes further than this to offer an overview of key issues and risks identified in the period covered by the report (often weekly in a longer study), a view of the status of the parking lot (loose ends not tied up) and a summary of key issues and concerns (which could of course include the pace of the study). Using a regular report like this can help the leader to maintain an overview of progress and key findings but is also a useful vehicle by which the sponsor or client can be kept informed and engaged on a regular basis, and warned of any difficulties that you might anticipate having to raise with them well in advance of this eventuality (especially if these are productivity issues involving their people). It also helps to keep the team more engaged because they are able to see the fruits of their work on a regular basis and can see that their management is being kept informed (and hasn't forgotten about them or the study). We'll also emphasise in Chapter 7 that producing regular reports will help you to assemble the final report as you go along by building up lists of key recommendations, assumptions and issues. An important aspect here is recommendations that the team consider to be associated with a significant risk (shown as high-risk recommendations in Appendix 19). The HAZOP progress report enables you to communicate such recommendations soon after they are generated, giving management the opportunity to address them at the earliest opportunity. When you proudly present your final report after the study, you do not want the reaction to be 'Why didn't you tell me about these issues at the time?' Some organisations insist that recommendations which carry a risk ranking above a certain threshold must be reported to management on the day on which they are identified. This helps managament to demonstrate their commitment to the study, and to process safety in general, by acting on perceived high-risk issues as soon as they have been informed.

Having to accelerate the pace of a study in order to achieve a required deadline carries the risk of adversely impacting its effectiveness, a phenomenon described by Hollnagel [4] as the 'efficiency thoroughness trade-off' or ETTO. There is only one way to avoid this, and that is to agree with the client or sponsor to reduce the scope of the study. Obviously, this may not be perceived as a suitable solution since it means the full scope of the HAZOP will not be completed, and additional arrangements will need to be made to complete it at a later date (with additional costs etc.). However, if the scope can be reduced by removing less hazardous parts of the process (utilities, services, water systems, vendor-supplied standard units for example) then this could be an acceptable compromise.

If reducing the scope of the study is not an option, then how can the pace be accelerated and what might be the negative impact of doing this? Box 6.1 summarises a number of possible options for trying to increase the pace of a study, all of which threaten its effectiveness.

Increasing the efficiency of the study meetings is often achieved by introducing a 5-minute rule; that after 5 minutes of discussion on any topic, for example the consequence of a hazardous scenario, then the discussion is brought to a conclusion and recorded (e.g. by a recommendation for more in-depth study) and the meeting moves on. This can be effective but carries the risk of reducing thoroughness, creating additional unnecessary recommendations (thereby creating more work outside the meeting) or frustrating team members who may feel they are unable to contribute fully or feel inhibited from speaking. Reducing the number of team members, reducing the number of deviations or increasing the size of nodes all present very real threats to the thoroughness of the study because each could easily lead to failing to identify hazardous scenarios. Studying parts of the process by analogy with other parts that have already been studied carries the risk of missing hazards in cases where the two processes may not be as similar as the team believes. Recording by exception could retain the thoroughness of the study but, as has been argued previously in Chapter 4 (Section 4.1.6) this could not be demonstrated in the final report in the form of high-quality worksheets. Switching to a

---

BOX 6.1 Speeding-up options.
- Increase meeting effectiveness e.g. 5-minute rule
- Reduce team size
- Reduce number of deviations
- Study by analogy
- Increase node size
- Record by exception
- Switch to What If?
- Stop projecting worksheets
- Change the leader

What If? Style of study is effectively abandoning the use of HAZOP, although there may be some justification in applying this simplified technique to parts of the process that are perceived to be less hazardous, or perhaps vendor-supplied units, provided the team has sufficient expertise (What If? can work well where the team has good knowledge of the equipment in question).

Some HAZOP practitioners in the USA have advocated [5] that projecting the HAZOP worksheets to the team throughout the meetings is detrimental to the pace and quality of a study on the basis that the team becomes excessively focused on 'word-smithing' what is being written rather than analysing the risk, thereby wasting meeting time. This book has argued strongly in favour of projecting worksheets as a means of promoting and maintaining engagement and would not advocate such an approach.

At this point it is worth considering the impact of any of the above measures on the motivation of the team. The importance of thorough, fully documented HAZOP has been communicated to them from the outset and their experience up to this time has lived up to this challenge. To then suddenly be told, 'Yes, I know I said it is important, and thanks for having worked with me in this way up to now, but we don't have the time anymore to do it like this' is likely to result in a loss of motivation which will itself further affect the study if it is performed in any way different to what has gone before.

The final point in Box 6.1 might be considered an attempt as humour, but it can be a realistic possibility and certainly has been tried. If you have lost significant time due to productivity issues, and you as leader have been unable to address these, then it may be appropriate to consider suggesting that you may not be the best person to continue to lead the study (of course you may have professional concerns in relation to diluting the quality of the study too). Your client or sponsor may already have had these thoughts too although, if they have, they should consider the initial negative impact on productivity of introducing a new leader with a new style for the team to get used to and a new group dynamic for the leader to try and develop.

In summary, most HAZOP studies exhibit some or all of the issues discussed here at some stage and the HAZOP leader must anticipate and manage them with the objective of minimising the impact on the effectiveness of the study. In the case of attendance, in most cases, this is successful. Maintaining energy is a core responsibility of the leader, but any changes induced by time pressure are likely to be detrimental to the effectiveness of the study. You should always keep in mind that there may be circumstances related to these issues or to others, in which you may feel that terminating the study is the least worst option.

6.3 Stopping the study

Terminating the HAZOP, or walking away from one as a leader, is always likely to be a difficult step to take: you may slow down an important project, antagonise your colleagues or your boss, or lose your contract as a third party HAZOP leader, so it needs to be considered carefully!

Unless the issue is one of extreme dysfunctional behaviour — threats of physical aggression, the use of insulting or abusive language or extreme emotional upset (possibly triggered by the threats or language) — the prime concern should be the effectiveness of the study. Are the circumstances such that the effectiveness of the study is being compromised to the extent that the final product will not meet the standards set by the organisation or accepted recognised good practice? Have I tried and failed to address the matter in collaboration with the client or sponsor of the study?

Perhaps the most common reason for terminating a study is that it is evident that the design is not of sufficient quality or completeness to apply the methodology effectively. The evidence usually takes the form of prolonged discussions that cannot be concluded due to the absence of information: a common example would be areas of piping and instrumentation diagrams (P&IDs) indicated 'to be finalised', or displaying design options or peppered with questions ('need for relief valve to be determined' for example). This will lead to the generation of high numbers of recommendations produced solely as a means of eliciting further information in order to clarify the design. Paucity or absence of supporting process safety information will produce the same effect as a result of the team being unable to answer basic questions relating to the design, for example questions relating to control philosophy or relief philosophy. Fortunately, these issues usually manifest themselves early in the HAZOP study, in which case the study is abandoned or postponed with minimal resource having been expended. If only specific parts of the process are as yet ill-defined, they can be removed from the scope in agreement with the sponsor or client, to be rescheduled when the design has sufficiently progressed.

Changes made by your client or sponsor to the scope of the study or the terms of reference may concern you to the extent that you feel the study should be terminated: the introduction of additional process units to the scope with no increase in time allowance; the removal of core team members to divert them to other projects or operating issues; a request to bring forward the completion of the study with no means identified to mitigate the impact on the study's quality; attendance falling short of previously agreed core team criteria.

In all the cases you will need to carefully justify the decision that the study should be terminated and offer objective evidence in support of this. Underlying these considerations will be the relationship with your sponsor or client (Will I be rewarded or punished?), your own professional reputation as a HAZOP leader (Will it be enhanced or will I become the 'awkward one' who's been known to walk away?) and your own professional ethics (I no longer think I should be associated with this). And when you begin to detect emerging issues that you think may escalate and excessively impact the effectiveness of the study, raise these at the earliest opportunity, offer to work with the sponsor or client to mitigate them, but at the same time develop some 'red lines' beyond which you are not prepared to go.

6.4 Assessing effectiveness

In Section 6.1 it was stated that we cannot objectively measure the effectiveness of a HAZOP study: we cannot state that we have identified 80%, 90% or even most of the hazardous scenarios in the process that has been studied because we don't know how many there might actually be. However, if we can't measure effectiveness objectively, we can assess how well the study was performed and imply from this how effective it was, and we can provide evidence to support the conclusion.

6.4.1 Quality assurance

The most common means of assessing how well a study has been performed is the retrospective application of a process of quality assurance (QA), which is often undertaken by another independent and experienced HAZOP leader, or perhaps a process safety technical authority or other role responsible for the HAZOP programme or for standards of deploying HAZOP within the organisation. The concept of quality is slightly different from that of effectiveness. QA aims to assess how rigorously the study has been conducted by examining the style and completeness of the final report and the associated process safety information used in the study. It does not look specifically for possible scenarios that may have been missed, which is how we have defined the concept of effectiveness, but effectiveness can be implied by the degree of rigour that the QA process uncovers. The purpose of such an undertaking is often not only to assess the quality (rigour) of a particular study but to seek opportunities for improvement in the organisation's application of the HAZOP methodology and management of HAZOP studies.

Over the years experienced HAZOP leaders and organisations have developed checklists to assist in HAZOP QA. An example of such a HAZOP QA checklist is presented in Appendix 20. To complete a checklist like this requires a relatively detailed inspection of the final HAZOP report, which should contain information or references to the majority of information required by the checklist. However, completion of the checklist also requires examination of the process safety information used in the study and therefore the assessor requires access to all the supporting information. The HAZOP report and supporting process safety information may well be extensive, but the assessor's job can be made easier if the HAZOP leader or recorder (or both) are available to assist in the assessment process.

Many questions in the QA checklist are of this type:

Were the terms of reference or charter documented and approved prior to the study?
Was the HAZOP documented in full to include the following: node descriptions, design intents, design and operating conditions etc.?

These are simple yes or no questions: was it done or not? There is no attempt to assess the quality of the terms of reference or the node descriptions. However, other questions in the checklist do attempt to assess the quality of the study at a deeper level and therefore go beyond strict QA, for example:

Are recommendations recorded according to "what-where-why-stand-alone" criteria and can they be interpreted without reference to the HAZOP worksheets?
Is there evidence that consequences were identified and taken to the ultimate unmitigated consequence?

These questions are much more probing and require a level of expertise in HAZOP to address them appropriately. If they are examined in some depth then the checklist will offer an experienced second opinion of the quality of the study.

6.4.2 Effectiveness questionnaire

Appendix 21 presents a slightly different approach in the form of a HAZOP effectiveness questionnaire. Although it has a lot in common with the QA checklist, there are some differences; for example questions that are relatively subjective and can only be answered by participants in the study, for example:

> *Did all team members contribute frequently to the study process?*
> *Were contributions open and constructive?*

Another difference is that it includes questions relating to the preparation for the study and follow-up of recommendations, so it is a little more wide-ranging than the QA checklist, which is more focused on the HAZOP report and its supporting documentation. Use of this questionnaire certainly requires the involvement of the HAZOP leader and preferably members of the team as well as those involved in commissioning the study and in following up the responses to recommendations. Although it is called an effectiveness questionnaire, like the QA checklist it does not look for possible missing scenarios and therefore does not address the way that we have defined effectiveness directly, but can be used to imply effectiveness by assessing how rigorously the methodology has been applied.

To some extent the approaches of the QA checklist and effectiveness questionnaire are complementary and could be further developed or customised in ways that suit a particular organisation. The important point is that they can be used to provide an independent assessment of the way in which the study has been managed, how well the methodology has been applied and the quality of the final report. This could constitute a form of implied assurance that the study has been effective in identifying hazardous scenarios and analysing them thoroughly.

6.4.3 How else could we assess effectiveness?

Most HAZOP leaders know intuitively how well the study they have just led has gone, but this is just a feeling: saying, 'Yeah, I think it went well' does not offer much insight into its effectiveness. But is there any evidence to which a HAZOP leader could point to support this assertion? Is there any other information not covered in the QA checklist or effectiveness questionnaire we have just discussed? What type of finding might we look out for that might suggest we have completed an effective study?

The first place to start is how it felt for you as leader and for your team. Were you allowed the time and resources that were agreed in the terms of reference? Did the study proceed as you planned it? Was the attendance of the team consistent and were the team dynamics positive throughout? What have team members said about their experience? Comments like 'We really learned a lot about the process and I learned a lot as an individual' are subjective but can be valuable evidence that you have had an effective study. Did team members actually have a positive experience and did they tell you about it without being prompted? Did their body language at the end of the study underpin their views, or were they just relieved that it was finally over? These informal signals can help to reinforce or reality-check your own views, so seek out how team members viewed their experiences and remember that negative comments offer you an opportunity to learn and improve as a leader.

Seeking more objective ways to assess the study, does it look like a complete and thorough study? Were all the nodes covered in sufficient detail and do the worksheets reflect this in terms of the quantity and quality of their content? Are hazardous scenarios identified for most deviations or are some of the worksheets 'thin-looking' — containing little detail or peppered with 'no hazards identified' and 'no significant consequences'? Seeing lots of 'no causes identified' can be even more concerning, since most deviations should be able to elicit some ideas for how they might be caused. One interesting question is whether there are protective devices shown on the P&IDs that are not incorporated in scenarios identified by the team. For example, if there is a pressure relief valve in a node but no high-pressure scenario identified in which it features as a safeguard then this could be evidence that a hazardous scenario has been missed. Such a check would be difficult to perform after the study except for smaller studies or, for larger ones, by means of random sampling of nodes. However, performing this check at the end of each node throughout the study is a simple and useful completeness check before moving from one node to the next; this quick check has been widely used by experienced HAZOP leaders and can be performed with the team. The pressure relief valve that you see on the P&ID but does not feature in a HAZOP scenario may have been provided as a regulatory requirement, for example as fire protection on a vessel, but it is still worth discussing and confirming this; noting the purpose and function of the device in the worksheets would be useful too. It could be useful to state in the final report, as a form of QA, that the purpose and function of every protective device in the design was identified, analysed and documented in the HAZOP.

Moving from protective devices that are incorporated in the design to those that are not, that is, situations where the team thinks additional protection is necessary, the next consideration in terms of effectiveness is the nature of the findings from the study. Have you produced recommendations that will be helpful to the organisation and reduce process risk if they are addressed? You are likely to want to highlight any recommendations that resulted from scenarios assessed as 'high risk' on the organisation's risk matrix, as we'll discuss in Chapter 7. If such recommendations suggest design weaknesses then this is likely to be of major assistance to the client or sponsor. An example would be the identification of a scenario with major accident potential in which the only safeguards are human actions such as response to process alarms; this would be a significant opportunity to reduce risk by means of active or passive layers of protection. In Chapter 7 we'll suggest that categorising recommendations in terms of those requiring additional active or passive protection (or significant cost, which is often the same) is a good way to grab management's attention in the final report.

More generally, risk ranking of recommendations before and after implementation of corrective action can be a useful way to highlight that the study has identified measurable risk reduction opportunities, although it is difficult to provide a frame of reference or benchmark for this (it should also be noted that conducting risk ranking before and after recommendations during the meetings will take up time and is probably best done after the study meetings).

The identification of hazardous scenarios that were not expected — that were a surprise to the team — could be another way to indicate that the study has been effective, especially in a study of an established operating unit. It's always interesting to hear an operator remark, 'I've worked on this plant for 15 years and we've never thought about this possibility'. It may be useful to draw attention to examples like this in the final report as evidence of effectiveness.

And finally, more generally, the study may have identified recurring themes or issues that have not been brought to the attention of management before, or may only have been brought to management's attention in isolated cases. Examples that have been seen in practice are equipment that has not been used ('It's never worked from the start') or not working correctly and subject to 'work-arounds', or a preponderance of control loops operating in manual mode, alarms shelved or operating procedures that have not been reviewed or updated for some time. This type of issue may not form a specific recommendation but would be worth mentioning as a general finding.

As stressed in the discussion of protective devices potentially missed by the study, evidence relating to the effectiveness of the study — key recommendations relating to 'missing' protective systems, unexpected scenarios or recurring issues etc. — are best monitored and recorded as the study proceeds. Some leaders maintain a 'key issues' list as a private note throughout the study, using it to build up a body of evidence that can be used to enhance the impact of the final report and, at the same time, provide evidence of the effectiveness of the study and promote its value to the client or sponsor.

There may not be an objective measure of your HAZOP's effectiveness, but as a leader it is good practice throughout the study to have in mind how effective the study feels and to look for examples that can be used in the final report to convince the reader that it has been an effective study that has provided value for the organisation.

The content of the final report and how it can be developed and presented is the subject of Chapter 7. The final report is your finished product: the HAZOP is not complete without it. And yet HAZOP leaders often don't pay it enough attention; some have been known to leave it to the recorder to write, others have taken months to produce the report and some have failed to produce a report at all (I was once hired to write a report for a study that I did not lead after the leader — a consultant in another company — failed to produce a report; it was a difficult assignment). You should recognise that a poorly presented final report can make the most effective HAZOP study look superficial, even if it wasn't.

References

[1] UK Health & Safety Executive, Quality Assurance of HAZOP, HSE Offshore Technology Report OTO 96 002, 1996.
[2] A. Hopkins, Lessons From Longford — The Esso Gas Plant Explosion, CHH Australia Ltd, 2000, pp. 28–29.

[3] Kenny P., Delta HAZOP: Revalidation and Focus on Major Accident Hazards, The Chemical Engineer, 2019.
[4] E. Hollnagel, The ETTO Principle: Efficiency Thoroughness Trade-off: Why Things That Go Right Sometimes Go Wrong, first ed., CRC Press, 2017.
[5] W. Bridges, R. Tew, Optimizing qualitative hazard evaluations for maximized brainstorming, in: Presentation at American Institute of Chemical Engineers, 2009 Spring National Meeting, 5th Global Congress on Process Safety, 43rd Annual Loss Prevention Symposium, Tampa, FL, 26—30 April 2009, Process Improvement Institute Inc, 2009.

CHAPTER 7

Develop your product

After a long and gruelling Hazard & Operability (HAZOP) study you can be physically, mentally and even emotionally drained, in which case the prospect of sitting down to compile a report that might represent weeks or even months of work is not particularly appealing. The start of the study might be forgotten in the dim and distant past; you are probably ready to forget about the intense experience you've just had (and possibly some of the people you've had to deal with); maybe you just want a rest or — more likely — another assignment awaits you. The urge to move on is likely to be considerable. And yet, in spite of all the work you've done, you are not finished yet; the HAZOP is not complete without a final report. You may have been feeding the recommendations into the orga-nisation's action tracking system as the study progressed — no bad thing, especially for recommendations requiring significant risk reduction — but the final report should offer more than just a list of recommendations and a set of HAZOP worksheets, particularly if you want to delight your cli-ent or organisation with a well-written report that will help them to deal with the overall conclusions and findings of the study and perhaps offer them insights that go beyond the list of recommendations.

In Chapter 7 we examine what should (and extra material that might) go into the final HAZOP report and how you can produce a high-quality report with a minimum of effort after the study meetings have been con-cluded. If that sounds like there's a 'silver bullet' or short-cut to a quick report then sorry, there isn't; it just means that the bulk of the work to compile the report is done during (and even some of it before) the study, so that finishing off the report in a few days following the study is not a demoralising and onerous task; it might even be a pleasure. We want to view the final report as the finished product that willdelight our customer; to do this we need to keep it in mind from the outset rather than leaving it to the very end, by which time we may well have forgotten the earlier parts of the study or even mislaid important material. If we can do this, we'll not only make the task of compiling the report easier, but it will also be of a higher quality and therefore a better product.

The HAZOP Leader's Handbook
DOI: https://doi.org/10.1016/B978-0-323-91726-1.00004-8

7.1 What should go in the report?

A template for a typical final report is provided in Appendix 22, together with some guidance on when each section can be completed, which we'll discuss later in this chapter. There is nothing particularly surprising here for reporting on a technical study: it needs to be a comprehensive, stand-alone document containing all information that was used, generated or referenced during the study; it needs to be presented professionally to its stakeholders in a way that important findings and issues are clearly presented and explained to management; all other findings need to be documented in an organised way that makes it as easy as possible for those who may need to consult the report in future, such as those involved with ongoing process modifications, follow-on studies such as Layers of Protection Analysis or repeat studies some years later.

All of this sounds pretty straightforward, but it can be incredibly difficult to consult an old HAZOP report and gain a good understanding of what was done, for manifold reasons... incomplete documentation, poorly recorded worksheets, unclear assumptions, uncertainty over who was involved, badly written recommendations and so on. Sometimes there is no information at all; the report has been lost or misplaced or is incomplete because the information was not stored in a single location (or not referenced adequately in the report) and has since disappeared. The discipline of compiling the report in a systematic way throughout the study will be discussed after we have emphasised three important aspects — the presentation of the main findings, the analysis of recommendations, and whether we should include an assessment of the quality of the study.

7.2 Presenting the key findings

Let's start with a simple assumption: nobody is going to read your report. The report from a 2—3 day HAZOP study, written using the template outlined in Appendix 22, will likely run to 40—50 pages; a study of several months duration could run into hundreds. Who's going to be interested enough or able to find the time to wade through that? The very size of it will put them off (even if the title 'HAZOP Report' doesn't). If you start from that premise and think about the stakeholders, or the people you would hope would be interested in your report (who are likely to be relatively senior members of management) then how can you make sure that your key findings reach the right eyes and ears? The only way you can do this using the report itself (without an accompanying verbal presentation) is to provide some attention-grabbing information in the executive summary, if possible on a single front page. You will probably not be able to get supporting explanations into that single page summary, but you can link your main points to a 'key issues, risks and recommendations summary' section in the main body of the report (even doc-link them, or provide a separate executive summary document for management containing only the summary and the key issues sections).

In the executive summary itself, you need to grab the attention sufficiently to have the reader thinking, 'I need to know more about this' and for them to dig down into the key issues section for further explanation. One possibility might be to present it as a newspaper would with a headline:

EXPOSED! MAJOR VULNERABILITY IN NEW PLANT DESIGN

It would be a brave HAZOP leader to try that one. However, if a significant concern was raised in the study then it should be highlighted. In addition to stating such concerns, it is useful to include a summary of the key recommendations — the top few or 'red risk' recommendations requiring senior management sign-off — as shown in Box 7.1, the bold text indicating document links to these places in the report.

BOX 7.1 Key recommendations in the executive summary.
This report contains a total of 87 recommendations, three of which are categorised as 'high risk':

- Recommendation 13 relates to the absence of sufficient overpressure protection in the Absorption Unit which could result in a multiple-fatality event.
- Recommendation 47 concerns the potential for a significant release of toxic reactant at start-up which could lead to a pollution incident with regulatory implications.
- Recommendation 63 addresses the risk of compressor damage in Unit 3 which could lead to significant production down-time.

Further details can be found in the **Recommendation Summary** and the full recommendations in **Appendix B.**

In addition to risk as a factor, senior management is also likely to be interested in recommendations that could be expensive to address, that may present a potential delay to the progress of a project if not addressed quickly, that may relate to a failure to meet a mandatory corporate requirement or to a potential regulatory compliance issue. These are things that can be presented as key risks or key issues.

It is also worth thinking further about your stakeholders — the different management roles that are likely to see the final report. Different stakeholders will have different interests; if you can present key findings in a way that speaks directly to your different stakeholders, then they are likely to be more interested. A project manager is likely to be interested primarily in the robustness of the design, its degree of completion and any potential impacts on cost: you can speak directly to these in the executive summary. An operations manager will be interested in the quality of operating procedures, any procedures that will be safety-critical and perhaps any implications for training. While an engineering or asset manager is likely to be concerned about the number of safety-critical devices or their performance, and the implications of HAZOP findings in relation to inspection, maintenance and testing programmes.

Performing an analysis of the study's recommendations is a prerequisite for developing a powerful executive summary, but remember that issues may emerge from the study that are not addressed directly by specific recommendations. For example, in a re-HAZOP you might find that throughout the study issues relating to poorly-maintained equipment or out-of-date operating instructions keep being brought up. In a HAZOP of a new design there might be numerous examples of technical details that have yet to be finalised. You may address some or more of these concerns with specific recommendations, but it may be appropriate to communicate them to management as recurring 'themes' of the study. If you do discuss 'themes' like this, it is important to consider the extent to which they may be viewed as an opinion. It is important to present your findings as those of the HAZOP team rather than your personal view; presenting them in a factual way can help. For example rather than saying 'one of the themes that emerged from the study was the back-log in operating procedure updating' it might be better to say 'the study identified more than 20 scenarios in which the operating instructions were not up-to-date'. We'll discuss the role of opinions again after looking at the analysis of recommendations.

7.3 Analysing recommendations

Unless the study is a relatively small one, in which case they can be presented as a section of the main body of the report, the full set of recommendations will form an appendix in the final report, where they can be presented as a list, possibly ranked by risk. Where there are a large number of recommendations it can be helpful to analyse them in a manner that you think will be helpful to the members of management that you hope will want to address them and summarise this analysis in the main body of the report.

We have already mentioned categorisation by risk and mentioned that recommendations with high cost or difficulty of implementation or that relate to corporate standards or regulatory compliance could be worthy of emphasis. There are many other ways in which recommendations can be 'sliced and diced' to present a useful analysis, depending on the audience for the final report, as discussed above. Some of these are summarised in Box 7.2.

An example of how they might be presented is shown in Fig. 7.1.

With this chart the HAZOP leader is drawing attention to the fact that almost half of the study's recommendations relate to risk reduction for the purposes of demonstrating that risks are as low as reasonably practicable (ALARP). This of course is the prime purpose of HAZOP, but the

BOX 7.2 Some possible ways to categorise recommendations.
Operating procedures
Maintenance procedures
Ongoing maintenance cost
(Potential) capital expenditure
Further investigation or analysis
Updating of plant records
As Low as Reasonably Practicable (ALARP) demonstration
Design checks and further ALARP assessment
Training and competence
Additional physical protection measures
Additional instrumentation
Inspection, maintenance and testing
P&ID and other document updates

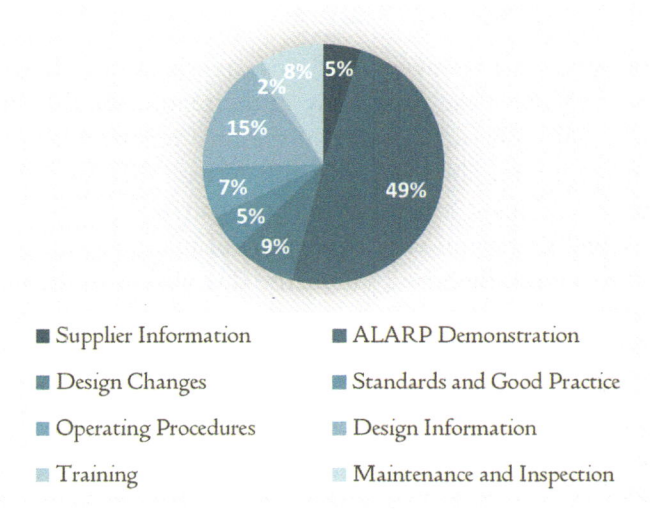

Figure 7.1 Summary of recommendations by type.

chart emphasises that the study has identified the need for a significant amount of risk reduction work. It also identifies a number of other areas requiring attention that could attract readers from different disciplines.

In discussing recommendations, we have mentioned three levels of presentation, which can be viewed as a hierarchy or pyramid, as shown in Fig. 7.2.

Remember you can document-link these places in your report to provide an easy way for readers to jump to and from them as they seek more detail, avoiding the need to 'trawl' through the report.

7.4 Should opinions relating to quality be included?

A section heading relating to an opinion of the quality, thoroughness or effectiveness of the study is not included in the report template shown in Appendix 22, for the simple reason that views are (unsurprisingly) split on the subject. Although it is likely that the HAZOP leader's name will feature prominently at the front of the report, it is important that the findings of the study are presented as those of the team as a whole. You will have explained this at the start of the study and perhaps many times again throughout the course of it, when facilitating discussions in which there are different views. As an experienced HAZOP leader you may have been disappointed in some aspects of the study compared to your

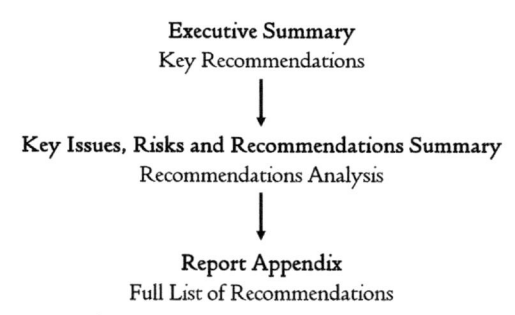

Figure 7.2 Presentation of recommendations in the final report.

experiences in other organisations or at other facilities: attendance may have been more patchy than you would have liked; process safety information may have been less extensive than you expected; you may have felt rushed towards the end of the study when pressure was on to complete it. As a result, you may feel that the quality of the study was not as high as you had intended. If this is the case, then from a professional standpoint you may feel the urge to make your point in the final report, since you are the author. The danger is, of course, that your team may not agree with you; it may have been fine according to their expectations, in which case you run the risk of being seen to be furthering your own opinions and not those of the team. Alternatively, if you hadn't raised any concerns during the study itself, then it may not look good to come along and raise them in writing when it's too late to do anything about it. One approach might be to discuss the quality of the study with the team and, if there is common agreement, record the team's view (if no concerns had been raised during the study itself you may still get criticism, of course). If you had raised concerns during the study, and these concerns had not been addressed to your satisfaction, then it would seem appropriate to record this in the report, albeit with the risk of a backlash from your client or sponsor anyway!

On a more progressive note, some organisations like to see an assessment of the quality of the study as a means to validate the effort that has been made, to record the views of the HAZOP leader as an independent expert that recognised good practice has been followed, or to help them to make improvements in the way the organisation applies the methodology in future studies. Some organisations require the completion of a quality assurance checklist like those described in Chapter 6, an example of which is presented as Appendix 20, sometimes by an independent HAZOP leader or process safety technical authority.

In conclusion, think about the quality and effectiveness of the study you've just led but take care in the way in which your conclusions are presented and, importantly, make sure there are no surprises! Remember that if, there were issues during the study, they should have been raised at the time.

7.5 Building up the report: before, during and after

The ideal point to reach at the end of the study meetings is that the work is almost over: there's just a final push of one or two days to complete the report. You don't want to be trying to assemble and compose a big report when you're already working on other projects or assignments, especially if it's another HAZOP! This might seem unrealistic for a report on a long-duration study but good planning, discipline and some extra work at the end of each day of the study can get you to the point that the only tasks that you need to do to complete the report are to complete the 'key issues, risks and recommendations section' and then write the executive summary.

We have already said that immediately the terms of reference are agreed and the preparation phase starts, we can put together a template for the final report, probably in a folder in which documents that will be included or referenced in the report can also be stored. That will enable us to give some focus to it from the outset, and of course provide a repository for the associated documentation as we build it up. Let's look at what can be done in relation to assembling the final report before, during, and after the HAZOP study.

Looking first at the preparation phase, the terms of reference provide us with a good starting point. Of course they will form an appendix in the final report, so we can complete this appendix immediately. They should also provide some other information for the main body of the report, for example the introduction, scope and objectives, the process description, the description of how the methodology is to be applied, the team membership and roles and the risk matrix. They may also provide additional material to be appended, for example the piping and instrumentation diagram (P&ID) list.

During the study itself, often the last thing the HAZOP leader and recorder want to do after 6 hours of meetings is another 1 or 2 hours of work, but unfortunately this is important for the smooth running of the study the following day and for building up the final report. Putting in this extra effort is easier if the study is taking place remotely and the leader and recorder are in a hotel in Aberdeen, Almaty or Atlanta, but where the leader and recorder are home-based there are inevitable family responsibilities, distractions from other work activities and social commitments that get in the way of evening work.

Some of the necessary daily maintenance activities that should be done at the end of each day are shown in Box 7.3.

Activities that will contribute to the development of the final report are shown in Box 7.4.

These tasks can be split between the HAZOP leader and recorder; typically the recorder would handle the worksheets, P&IDs, additional materials and glossary, leaving the leader to focus on developing recommendations and adding to or refining lists of assumptions and key issues. Although this would appear to be an arduous task, it is particularly important from a quality perspective. The quality of recommendations and communication of key issues are the most important aspects of the final report, your product. The day on which you identify the recommendations or issues is the time when these are freshest in your mind and therefore this is the best time to record them. In particular, crafting a recommendation requires an explanation of the context for the recommendation and the risk associated with it that may well be forgotten by the end of the study if they are not fully captured at the time.

That brings us back to the end of the study and what remains to be done to complete the report: analysis of recommendations, the executive summary and some tidying up; hopefully no time-consuming editing of worksheets, writing of recommendations, remembering who was there, scanning drawings and searching for reference documents.

By building up the final report from the beginning we've reduced the post-study workload and hopefully produced a product of high quality

---

BOX 7.3 Daily maintenance activities.

Updating the parking lot and action/issues list

Updating the attendance and session logs

Escalating issues such as newly discovered high-risk scenarios to management

Updating trackers to monitor progress

Checking drawings and node descriptions for the following day

> ### BOX 7.4 Daily activities that develop the final report.
>
> Reviewing the day's worksheets for spelling, grammar, completeness and quality and circulating to the team
>
> Developing recommendations in what where why stand-alone format and circulating to the team
>
> Updating the assumptions list
>
> Updating the 'key issues' list
>
> Signing-off, scanning and filing completed P&IDs
>
> Identifying and filing additional materials to be appended to or referenced in the final report
>
> Adding to the glossary of terms and acronyms.

that is presented in an attractive way to the client or sponsor. Delighting the customer is perhaps too strong a word for it, but the least we can do is to maximise the likelihood that the findings from the study will be addressed by presenting them comprehensively and, at the same time, in a way that highlights the key issues.

We've travelled a long way from the conception of the study through its planning, its technical execution and the facilitation skills necessary to maximise its effectiveness, looking for evidence of its effectiveness and, finally, the production of a final report that has a real impact and will provide management with a set of recommendations that, when addressed, will reduce risk and enhance the operability of the process. We have discussed processes and tools that are available to assist the HAZOP leader in their formidable task but we have also emphasised the importance of the skills and experience of the leader. In Chapter 8 we'll conclude with some of the things that you, whether an aspiring or practicing HAZOP leader, can do yourself to become more effective at the planning, execution and reporting of your studies, while at the same time promoting best practice and maintaining the reputation of the methodology.

CHAPTER 8

How to be a better HAZOP leader

I said the outset that the prime purpose of this book is to provide guidance specific to Hazard & Operability (HAZOP) leaders to help you to maximise the effectiveness of your studies, thereby getting the most benefit from the methodology, promoting consistency and rigour in its application and sustaining its well-earned reputation to attract the next generation of HAZOP leaders. I hope that you have discovered some useful tips, techniques and materials that you can utilise in your efforts to develop as a HAZOP leader. After all, the ideas for the book came from observing and talking to a large number of experienced leaders over many years.

We'll end the book by drawing together, as a form of summary, 10 things you can do to become a better HAZOP leader.

1. Try to get involved at the earliest possible stage of the HAZOP that you have been asked to lead, preferably at its conception, and then drive the planning and preparation process so that things are set up for you as you want them to be at the start of the study. You will go into the study fully prepared and focused on getting the team into 'performing' mode as quickly as possible.

2. Agree on terms of reference for the study — your contract — with the study's sponsor or your client and aim to make sure that they give you adequate time and resources to do your job well. The terms of reference will act as your insurance policy if things don't work out as well as you expect.

3. Value your team. Get involved in their selection if you can; if you can't then get to know them before the study begins. Motivate them all to get and stay involved. Respect their knowledge and experience and make their time in the meetings beneficial and enjoyable.

4. Pay special attention to selecting the best recorder you can find, coach them in your expectations and forrge a close-knit partnership with them. The recorder role is so important to a successful study. A good recorder will free you to focus fully on your role as a facilitator and will give confidence to the team.

5. Undertake some professional facilitation training and apply it in your studies! Don't just chair them like a business meeting but be enthusiastic and let your enthusiasm encourage involvement and creativity throughout the team.

6. Strive to apply the 'golden rules' as thoroughly as you can in identifying and analysing hazardous scenarios, and pay particular attention to crafting recommendations in the 'what-where-why-stand-alone' format; the recommendations are the lasting legacy of the team's work, so take a personal lead in formulating them for maximum effect.

7. Never forget the lessons from your operating experience. Be sceptical but never cynical: equipment fails; people don't always follow procedures (which are often out-of-date); people often don't spot problems developing; people don't always respond appropriately to alarms; the process is often in a constant state of change and modification; new operating issues are identified regularly. Make sure you have a set of examples from your career and well-known incidents to educate the team if necessary.

8. Think about and assess the effectiveness with which your team is identifying and analysing hazardous scenarios and look to maximise it. Gather evidence of that effectiveness and incorporate it in the final report.

9. Develop the final report as the study progresses; spend quality time on writing it, make sure you know who is going to receive it and make sure it grabs their attention so that they do read it.

10. Be prepared to say no to leading a study if you are unable to secure the time and resources you need. If these are retracted during a study and its quality starts to be affected, raise your concerns, formulate some 'red lines' and be prepared to terminate or leave a study if it compromises the organisation's or your own professional standards. You have a duty to safeguard the reputation of the methodology.

I hope that you have identified many other tips that will work for you and that the book has raised your enthusiasm for the role of HAZOP leader.

Three other reasons for this book were stated in the introduction and it is worth closing by reminding you of your responsibilities in respect of these. First, preserve the integrity of the methodology by resisting the ever-increasing pressure within organisations to cut back on study time. Second, promote the application of current best practice, which in the view of myself and other leaders has moved ahead of that described in the HAZOP literature, but is documented in this book to as large an extent as possible. Third, increase the emphasis on facilitation skills as a core competence for HAZOP leaders. And finally... remember to say to yourself when you wake up, 'it's HAZOP today and I'm up for it!' and then go and show your team that you are.

Appendices

Appendix 1 HAZOP practitioner assessment protocol

HAZOP PRACTITIONER ASSESSMENT SUMMARY							
Name:							
Assessor:							
Date:							
Aspect		1	2	3	4	5	Evidence
Qualifications	First degree or equivalent	■	■	■	■	■	
Depth of experience	Engineering and technical experience, including process design and operations	■	■	■	■	■	
	Completion of recognised training for HAZOP leaders						
	Six months in the apprentice role						
	Chaired at least three HAZOP studies under supervision and scribed at least one study						
Width of experience	Good awareness of other hazard study stages and when they are applied, sufficient to advise project managers. Able to lead FMEA and checklist studies. Awareness of when to apply other specialist tools, e.g. hazard and consequence assessment. Spending one third of time on HSE-related issues	■	■	■	■	■	

(Continued)

HAZOP PRACTITIONER ASSESSMENT SUMMARY

Professional standards	Ability to make professional judgments in testing circumstances where technical knowledge may be limited	☐	☐	☐	☐	☐	
Regulations	Knowledge of relevant regulatory requirements	☐	☐	☐	☐	☐	
Engineering standards	Knowledge of national and international standards and guidelines relevant to process safety	☐	☐	☐	☐	☐	
HAZOP methodology and recognised good practice	Evidence of application of RGP in HAZOP reports	☐	☐	☐	☐	☐	
Personal competencies and facilitation skills	Have the required personal competencies for self-management, communication and ability to influence. Suitability to mentor other trainees	☐	☐	☐	☐	☐	

Key: 1 = very weak, 2 = weak, 3 = adequate, 4 = strong, 5 = very strong.

HAZOP PRACTITIONER ASSESSMENT PROTOCOL

Name:	
Assessor:	
Date:	

Area	Criterion	Evidence	Further Questions?
Qualifications	Qualification to first degree level in an engineering discipline.		

(*Continued*)

HAZOP PRACTITIONER ASSESSMENT PROTOCOL

Depth of experience	Attendance at a recognised HAZOP leaders training course with subsequent assessment		
	Two years in an engineering or technical function. Experience of process design and operations.		
	Chaired at least three HAZOP studies under supervision and scribed at least one study.		
	Participation in at least one major capital project HAZOP study as a full-time member.		
	Apprentice HAZOP leader for longer than 6 months.		
	At least one third of time spent on HAZOP or related process safety activities.		
Width of experience	Experience of hazard reviews at all stages of a project, e.g. the 6-stage process.		
	Understanding of the timing of HAZOP and other hazard reviews in the project life cycle.		
	Understanding of corporate and business unit requirements for HAZOP and other risk reviews		

(Continued)

HAZOP PRACTITIONER ASSESSMENT PROTOCOL

	Awareness of corporate and business unit process safety management and technical engineering standards, e.g. relief systems and fire protection.		
	Experience of the use of checklists, what if and FMEA.		
	Understanding and experience of the application of HAZOP to batch processes, PES, electrical systems, manual operations and operating procedures.		
	Understanding and experience of the application of HAZOP to (vendor-supplied) packaged units.		
	Understanding of other specialist risk assessment tools and when they are appropriate, e.g. LOPA, HAZAN, human error assessment and consequence analysis. Sufficient to advise project managers		
	Experience and understanding of risk ranking, preferably using the corporate model.		

(*Continued*)

HAZOP PRACTITIONER ASSESSMENT PROTOCOL

Professional standards	Ability to make professional safety judgments in demanding circumstances and seek support and advice where technical knowledge may be limited.		
Regulations	Knowledge of relevant safety and environmental requirements.		
Engineering standards	Knowledge of relevant national and international engineering standards, e.g. API, NFPA and ISO.		
HAZOP methodology and recognised good practice	Ability to estimate the time required to complete a HAZOP study.		
	Understanding of the importance of scope definition and the definition of objectives and deliverables, e.g. recording policy, consideration of hazard vs operability, safety vs HSE, major hazards vs all hazards, assumptions relating to DCS or PES reliability, vendor packages, action recording protocol, risk ranking protocol, report format.		
	Understanding of the minimum requirements for the composition of the HAZOP team.		

(*Continued*)

HAZOP PRACTITIONER ASSESSMENT PROTOCOL

Understanding of the minimum information required to complete an effective study (frozen and accurate P&IDs, process flow diagram with operating conditions, equipment design parameters, etc.).		
Understanding of the role of the facilitator (facilitator to the team in ensuring identification of all relevant hazards with the optimum use of team time).		
Understanding of the pitfalls associated with HAZOP, e.g. inadequate design or information, team composition, limiting assessment of consequences, boredom, redesign.		
Ability to frame recommendations according to "SMART" principles.		
Awareness and experience of HAZOP recording software, e.g. PHA-Pro.		
Understanding of the risk ranking process and the ability to facilitate risk ranking of events and/or recommendations quickly using a risk matrix.		

(*Continued*)

HAZOP PRACTITIONER ASSESSMENT PROTOCOL

Personal competencies and facilitation skills	Ability to involve all team members.		
	Ability to resolve disputes.		
	Ability to maintain the pace of the study.		
	Ability to refuse to start or to terminate in the absence of information, team members, wrong time in project, poor design, etc.		
	Concern for standards and thoroughness in recording the study and in the quality of the final report.		

Appendix 2 HAZOP preparation guide

Phase	Consideration	Guidance
Preliminary *Is this the right thing to be doing and are the expectations realistic?*	What are the objectives of the proposed study or programme?	The motivation for undertaking a study may influence the expectations of the client or sponsor and the resources at their disposal. Possible motivations are routine compliance with corporate process safety management system requirements (e.g. revalidation or project management requirements) regulatory pressure (e.g. response to an incident) or management concerns in relation to process safety or operability issues. It is useful to understand the level of commitment to a thorough HAZOP study and the level of commitment to addressing operability aspects as well as process hazards.

(Continued)

Phase	Consideration	Guidance
	Is HAZOP the most appropriate technique to address these objectives?	Other hazard identification techniques such as HAZID or FMEA may be more appropriate in addressing the client or sponsor's needs.
	Does the client or sponsor have a good understanding of the time and resources required for successful HAZOP and the nature of the output?	Early indications from the client of the time and resources they are expecting to allocate to achieve the objective may indicate that they may underestimate what is involved or not understand the HAZOP technique. It is important to manage these expectations or back off if the risk of failure seems significant and you believe you may be in danger of being "set up to fail."
	Does the client or sponsor have the authority and resources to commission and execute the study and follow up the recommendations?	The study could stall or fail in its objectives if the commitment and resources are not made available.
	For a new facility or modification, is the definition of the design sufficiently detailed to enable HAZOP to be applied effectively?	Process design should be more than 90% complete and P&IDs almost fully developed. Applying HAZOP too early risks it turning into a design review and/or generating very large numbers of recommendations.
Scoping *Can we make a realistic plan?*	What are the boundaries of the study and how many P&IDs will be covered by the study?	The extent of the study should be defined in terms of interfaces with other facilities, and any equipment or units to be excluded from the scope should be identified. The P&ID listing is important information for estimating the duration of the study.

(Continued)

Phase	Consideration	Guidance
	Has the facility been subjected to HAZOP before and, if so, what is the quality of the output and were the recommendations tracked to completion?	If the quality of the previous study was poor, or significant changes have been made, then redo may be a better option than revalidation.
	What is the extent and quality of the process safety information that is available?	P&IDs in particular should be up-to-date and accurate. It may be necessary to review and update P&IDs as a preparatory phase of the study.
	Is the study of a size that could warrant appointing a dedicated HAZOP Coordinator from the client organisation to act as a contact point for the organisation and execution of the study?	For larger studies the provision and control of process safety information, coordination of team members (and possible vendor representatives), meeting facilities and other logistical tasks represent a significant workload.
	Does the client organisation have a corporate standard for the conduct of HAZOP?	If this is the case, care will be required to ensure that the Terms of Reference address all corporate requirements.
	What expectations does the client have in relation to the recording of the HAZOP?	There are three aspects to this: the first is whether proprietary software is to be used (in which case suitable licences may need to be purchased); the second is whether a dedicated recorder is to be used, in which case the

(Continued)

Phase	Consideration	Guidance
		pedigree of this individual will need to be assessed and suitable coaching and software familiarisation organised. If a dedicated recorder will not be provided, this will have a significant impact on the time estimate for the study. The third aspect is the style of recording; full recording will increase the duration, but is the recommended style.
	How long is the study going to take?	It is important that the client's expectations are managed from the outset. If there is a big difference between your estimate and the client's expectations then this will need to be debated transparently in relation to scope, depth (e.g. number of guide words), node size and recording style. Include preparation and report writing time in the estimate.
Terms of Reference *Can I develop and agree a robust contract with the client or sponsor that reduces the risk of issues and gives me a way to address them?*	Does the client organisation have a standard practice for developing and recording Terms of Reference?	If the client organisation does not employ the concept of Terms of Reference, it is recommended to push for written agreement covering scope, methodology, process safety information to be provided, personnel to be involved and the schedule and deliverables.
	What problems can I anticipate?	Try to build in mechanisms for preventing or addressing potential problems into the Terms of Reference. Common threats are: • Poor scope definition—include a list of P&IDs and their revisions in the ToR;

(Continued)

Phase	Consideration	Guidance
		• Poor quality of design and/or P&IDs—ensure P&IDs are "approved for HAZOP"; • Poor availability and/or quality of process safety information—specify required documentation and its means of retrieval; • Poor attendance – specify attendants by name, identify the core team (no show—no meeting) and seek named deputies; • "Burn out"—agree limits for days per week and hours of study per day. Include requirements for progress reviews to ensure the mechanism is in place to address problems quickly.
	What are going to be my "red lines"?	What things could happen that might require me to stop the study and revert to the client? Examples would be P&ID or design quality, process safety information, attendance issues (too many or too few), difficult behaviours or substandard meeting facilities.
	Are the Terms of Reference agreed by a client representative with sufficient authority to address issues if and when they arise?	It is important to be able to address issues proactively and quickly without escalating them up the organisation for resolution.
Execution	Can I undertake the maximum amount of prework before the event?	Allowing the HAZOP team to focus on the task is important for motivation and maintaining commitment.

(*Continued*)

Phase	Consideration	Guidance
		Aim to complete node identification and marking, node descriptions and software pre-population. Doing these activities in the HAZOP meetings can lead to unnecessary frustration and loss of motivation.
	Can I secure the process engineer from the facility to assist in node identification?	This will produce better node definitions in less time.
	Can I secure time with the recorder before the study?	This is important for ensuring familiarity with the recording medium, agreeing expectations in terms of recording style and establishing some rapport in advance of the start of the event.
	How can I make the meeting environment as conducive as possible to good team-working?	Try to influence the choice of a location that will be convenient and comfortable for the team in relation to travel, parking, comfort, IT, quality of refreshments and space for relaxation.
	How can I start the study in a dynamic and positive style?	Prepare an agenda for Day 1, a short introductory presentation and draft ground rules. Aim to engage the team right from the outset. Make sure that all the required process safety information is in the room and that refreshments are provided.
Reporting	What information can I get in advance to start writing the final report?	Develop a final report template and start to populate it as early as possible, e.g. with the introduction, process description, methodology description, Terms of Reference and risk matrix.
	Do I have the mechanisms in place to develop	Examples include scanning of completed P&IDs, worksheet review and agreement and

(Continued)

Phase	Consideration	Guidance
	the report as the study proceeds?	development of recommendation sheets on a routine basis throughout the study.

Appendix 3 Typical structured what if? or SWIFT checklist

Structured What If? Checklist (SWIFT)

Equipment failure or maloperation

Valve erroneously closed, opened, leaks or passes

Trip fails

Check valve fails

Overpressure protection

Loss of containment

Other

Loss of utility

Power, e.g. pumps, compressors, air coolers, MOVs

Steam, e.g. heaters, turbines

Cooling fluid, e.g. coolers, rotating machinery

Instrument air, e.g. control loops, control valves

Contamination

Tube leakage

Wrong additive

Air ingress

Other

Maintenance

Will maintenance be carried out while plant is running?

How will isolations made for maintenance affect the system?

How will isolations for maintenance be achieved?

Other

(*Continued*)

Structured What If? Checklist (SWIFT)

Corrosion or erosion

Corrosion or erosive fluid service

Dead legs

Other

Hazardous materials

Chemicals/hydrocarbons

Clinical waste

Food waste

Sewage (treated or raw)

Radon

Other

Manual labour

Moving machinery/crush hazard

Equipment guards (mechanical)

Low or high equipment surface temperatures

Working at height

Repetitive tasks

Other

Environment

Heat

Dust

Sandstorm

Heavy rain or flood

Other

Human factors

Driving

Human machine interface (instrument visibility, controls accessibility)

(*Continued*)

Structured What If? Checklist (SWIFT)

Work area layout/access routes

Emergency response, e.g. escape, mustering, communication, alarms

Other

Appendix 4 Terms of Reference

Terms of Reference contents

Title	Content
Cover	Subject Signatures of sponsor and HAZOP leader Date of agreement
Background	Historical, technological, geographical
Objectives	
Scope	Description of the process Systems and facilities to be studied Boundaries and interfaces Systems and facilities to be excluded List of P&IDs to be studied (reference appendix)
Methodology	Type of study (continuous, batch, procedural or mixture) Modes of operation to be studied List of deviations to be studied Node planning guidelines Requirements for developing and recording causes, consequences, risk ranking and recommendations Use of risk matrix P&ID management protocol "Parking lot" protocol
Process safety information	Information required in the meeting Information to be available electronically
Personnel	Leader, recorder, team including disciplines Core team members and deputies Other part-time members
Schedule and deliverables	Plan of execution (timings, location, etc.) Progress reporting requirements and highlighting of significant issues

(Continued)

Terms of Reference contents

Title	Content
	Final reporting requirements Process for handling recommendations
Report content and distribution	Required contents Appended information Main recipient Other recipients

Appendix 5 Process safety information

Essential (core) documentation

Process description and process chemistry	
Facility plot plan/unit layout drawings	
P&IDs "Approved for HAZOP"	Vendor packages if within scope Piping class specifications Materials of construction
PFDs	Heat and material balances Inventories Safe upper and lower operating limits, operating envelopes
Previous HAZID, HAZOP or LOPA reports	
Control, alarm and trip information	Alarm and trip settings Control system philosophy and description Interlock/trip activation and response descriptions Shutdown matrices (cause and effect diagrams) ESD system functions
Pressure relief, flare, vent and de-pressuring information	Relief valve data sheets Scenarios considered for sizing of the devices Flare/disposal systems design and sizing information
Process sequence, for batch operations	

(Continued)

Essential (core) documentation

Changes to design since the last HAZOP	For re-HAZOP
Operating procedures	Start-up, operating, shutdown, emergency
Previous incident reports	For re-HAZOP

Additional information

Corrosion control guidelines and corrosion and materials diagrams
Emergency de-pressuring system functions
Pump and compressor operating curves and dead head pressures
Instrumentation data sheets, including control valves, orifices, throttling valves and regulators
Valve capacities—particularly important for gas breakthrough
Fire protection design philosophy and basis
Inspection and testing results, maintenance records, operational history and current condition of process equipment
General arrangement and elevation drawings, including electrical area classification and drainage
Vessel inventories
Operations and maintenance philosophy
Commissioning procedures
Maintenance procedures
Material Safety Data Sheets (MSDS)
Previous risk assessment, in particular, any consequence modelling
Electrical loop diagrams
Ventilation system design
Design codes and standards employed

Appendix 6 Node selection guidance

Type of equipment	Node selection guidance
Vessels	Major vessels should form a separate node, e.g. separators, flare drums, storage tanks, reactors. A "major" vessel can be considered as one that contains a fluid with a relatively high hazard potential and has several branches (feeds and outlets). Local vents, level bridles and local drains should be included.

(Continued)

Type of equipment	Node selection guidance
	A "minor" vessel can often be included within the scope of a node which includes the piping either side of it. "Minor" can be considered as having a relatively low hazard potential, e.g. a small knock-out drum.
Process lines	For each feed line into a major vessel, follow it back to its source. If this is at the start or boundary of the study then start the node there (the "interface"). If it comes from an upstream vessel, consider starting node at the upstream vessel outlet. If there is a phase change, then consider defining the node so that it contains a single phase. Consider number of control loops and branches in the process line. If there are more than two control loops and several branches, e.g. to drains, vents, flare systems, consider fragmenting the node; more than two control loops may make the node too large, requiring extensive discussions. Typically, a node with a single control loop is likely to be manageable chunk. Remember, "…the guide word should apply uniformly throughout the node." After reviewing all the feed lines, conduct a similar process for the outlet lines.
Pumps	These are typically included within a piping node, to include any kick-backs/recycle lines.
Heat exchangers	Aim to include the temperature control loop within the node. Utility and process sides can usually be considered each as one node. Identify the boundaries on the utilities lines as the first upstream manual isolation valves.
Compressors	Include the process side within piping node(s) in and out, but identify separate nodes for seal systems and utilities. Multi-stage compressors should be split up stage by stage.
Loading and unloading	Ideally tankers should be included on the P&ID to clearly show the connection and loading or unloading arrangements, tanker fittings, relief streams and any vapour recovery or vent lines. The boundaries of the node should include the tanker and line to the entry point on the vessel or exit point from the vessel. If vapour return involved consider splitting this into a separate node. If loading or unloading is a hazardous operation with significant human intervention, consider conducting a procedural or human HAZOP for the activity.

Appendix 7 HAZOP deviations

This is a list of deviations that have been used in HAZOP. Select from it carefully—not all of these are correct deviations!

No/less flow
More flow
Reverse/misdirected flow
Less pressure
More pressure
Less temperature
More temperature
Less viscosity
More viscosity
High level
Low level
High pH
Low pH
Loss of/less agitation
More agitation
Different phase
Different composition
Different concentration
Contamination
Oxygen ingress
Relief
Mixing/reaction
Different operation
Equipment
Instrumentation
Valve fail positions
Redundancy
Corrosion/erosion
Sampling
Loss of containment
Start–up
Shutdown
Emergency shutdown
Service failure
Maintenance

Decommissioning
Static
Fire/explosion
Safety
Training
Environmental

This list is mainly derived from organisations employing reactive chemistry:

Mix up of components
Contamination of components
Dosing quantity too high
Dosing quantity too low
Mass flow too high
Mass flow too low
Too early (point of time)
Too late (point of time)
Wrong sequence
Residence time too short/step done too quickly
Residence time too long/step done too slowly
Temperature too high
Temperature too low
Pressure too high
Pressure too low
Insufficient mixing
Wrong conveying route
Wrong conveying direction
Outward leak
Internal leak
Leaking valve
Build-up of explosive mixture
Generation of ignition source
Wrong proportions of substances
Wrong particle size
Wrong state of aggregation
Concentration too high
Concentration too low
Catalytic effects
Inhibitory effects

Catalyst activity too low
Viscosity too high
Viscosity too low
pH too low
pH too high
Corrosive/abrasive effects
Solidification/incrustation/sedimentation
Choking/sticking/deposition build-up
Condensation/crystallising
Degassing/foaming
Flocculation
Incompatibility of substances
Level too high
Level too low
Decomposition
Wrong phase
Phase change
Return flow
Siphoning off
Static charging
Loss of cooling
Cooling too strong
Loss of heating
Heating too high
Prevented liquid expansion
Loss of agitation
Agitator too slow
Loss of vacuum
Intake of air
Loss of inerting
Pump failure
Wrong impeller
Agitator breakage
Filter breakage
Break of column trays
Gasket leakage
External corrosion
Corrosion under insulation

Appendix 8 P&ID management protocol

Key principles

- SINGLE MASTER SET
- NO LOOSE DRAWINGS
- HAZOP TEAM ROLES OWN THEIR SETS

Client establishes a Master List of P&IDs from the HAZOP scope and incorporates into the Terms of Reference including vendor drawings and all revision numbers.

Print one full set of P&IDS for node marking in A0, A1 (preferred) or A2.

HAZOP Leader and Asset Process Engineer mark nodes on P&IDs to form the **Master Set**.

Create an A3 set of marked-up P&IDs in a binder for each HAZOP role (not individual member if different people may represent one or more roles).

Each HAZOP role is responsible for maintaining this set complete and intact and handing this set over to successors/replacements when required.

HAZOP Leader to stamp each P&ID in the master set as "Approved for HAZOP" and "HAZOP Master" and place in a binder.

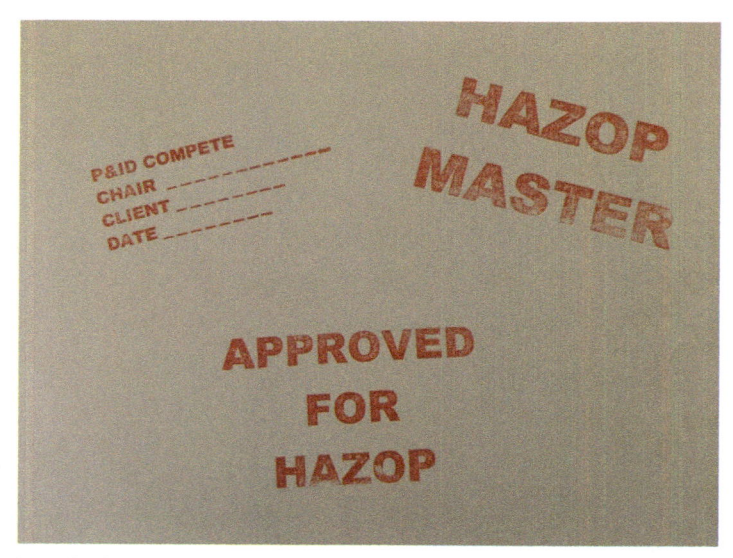

Figure A8.1 P&ID stamps.

HAZOP Leader is the custodian of the master set.

There is only ever one master version!

If any changes to nodes are made, create a new master, re-mark, copy and replace unless the change can be clearly shown without confusing the P&ID or any other node mark-ups.

Record the node completion dates, relevant recommendations references and required drawing updates on each completed P&ID and sign off by HAZOP Leader and Client Representative.

If new P&IDs are uncovered, add to the Master List, revise Terms of Reference, and repeat this process, destroying any replaced drawing.

Appendix 9 HAZOP event checklist

Category	Item	Notes or status
Technical	Terms of reference authorised with copies for each Team Member	
	Node identification and master P&ID mark-ups	
	Protocol for management of P&IDs	
	Protocol for management of the "Parking Lot"	
	Software population: administrative sections (descriptive sections, meeting dates, team members); nodes and node descriptions; deviations	
Administrative	Meeting room booked	
	Meeting notices sent out (arrive 15' before start time) with directions	
	Building security access and parking	
Introductory	Introductory presentation	
	Attendance register	
	Ground rules proposal	
	Project representatives or client/sponsor attendance (if required)	

(*Continued*)

Category	Item	Notes or status
Facilities	Room size and layout	
	Audio-visual equipment (consider two projectors)	
	Refreshments prior and during	
	Flip charts and pens	
	Highlight pens	
	Blutac or masking tape	
	Skype, Zoom or other means for video conferencing	
	Name cards (tents)	
	Emergency arrangements	
	Wi-fi access	
Documentation	Hard copy process safety information for room	
	IT links for access to electronic process safety information	
	Master P&IDs, A0 or A1, stamped "Approved for HAZOP"	
	Sets of A3 P&IDs for each HAZOP team role	
Other	P&ID rubber stamps ("Approved," "Master" "Completion")	
	Magnifying glass(es)	

Appendix 10 Modes of loss and potential causes

From CCPS Guidelines for Chemical Process Quantified Risk Assessment, 2000, Appendix A

Mode of loss	Potential causes	
Containment lost via an "open-end" route to atmosphere	Due to genuine process relief or dumping	
	Due to maloperation or equipment in service, e.g. spurious relief valve operation or rupture disk failure	
	Due to operator error, e.g. drain or vent left open, misrouting of materials, tank overfilled, unit opened under pressure	
Containment failure under design operating conditions due to imperfections in the equipment	Imperfections arising prior to commissioning and not detected before start-up (due to poor inspection and testing procedures)	Equipment inadequately designed for proposed duty, e.g. wrong materials specified, pressure ratings of vessels or pipe work inadequate and temperature ratings inadequate
		Defects arising during manufacture, e.g. wrong materials used, poor workmanship and poor quality control
		Equipment damage or deterioration in transit or during storage
		Defects arising during construction, e.g. welding defects, misalignment and wrong gaskets fitted
	Imperfections due to equipment deterioration in service and not detected before the effect becomes significant (due to inadequate monitoring where deterioration is gradual)	Normal wear and tear on pump or agitator seals, valve packings, flange gaskets, etc.
		Internal and/or external corrosion, including stress corrosion cracking
		Erosion or thinning
		Metal fatigue or vibration effects
		Previous periods of gross maloperation, e.g. furnace operation at above design tube skin temperature ("creep")
		Hydrogen embrittlement
	Imperfections arising from routine maintenance or minor modifications not carried out correctly—poor workmanship, wrong materials, etc.	

Mode of loss	Potential causes			
Containment failure under design operating conditions due to external agencies	Impact damage such as by cranes, road vehicles, excavators and machinery associated with the process.			
	Damage by confined explosions due to accumulation and ignition of flammable mixtures arising from small process leaks, e.g. gas build-up in analyser houses, enclosed drains, around submerged tanks			
	Settlement of structural supports due to geological or climatic factors or failure of structural supports due to corrosion, etc.			
	Damage to tank trucks, rail cars, containers, etc. during transport of materials on or off site.			
	Fire exposure			
	Blast effects from a nearby explosion (unconfined vapour cloud explosion, bursting vessel, etc.) such as blast overpressure, projectiles, structural damage and domino effects.			
	Natural events such as windstorms, earthquakes, floods and lightning.			
Containment failure due to deviations in plant conditions beyond the design limits	Overpressure of equipment	Due to connected pressure source	Gas pressure source	(1) Gas breakthrough into downstream low pressure equipment due to failure of a pressure or level controller, isolation valve opened in error, etc. (2) Pressurised backflow into low pressure equipment, e.g. due to compressor failure.
			Liquid pressure source	(1) Pumping of blocked-in gas spaces (2) Hydraulic overpressuring due to blocked-in condition downstream (3) Excessive surge or hammer, such as by sudden valve closure on liquid transfer line

Mode of loss	Potential causes			
		Due to rising process temperature	Loss of cooling	(1) Loss coolant flow, e.g. to a reactor cooler or condenser (2) Elevated coolant temperature, e.g. loss of cooling water fans.
			Excessive heat input (thermal)	Heater control faults such as on steam or oil heated systems
			Excessive heat generation (chemical)	(1) Reaction runaway, e.g. due to loss of reaction diluent, high feed to inadequate mixing or temporary loss of reaction leading to runaway (2) Exotherm due to ingress of catalytic impurities (3) Exotherm due to mixing of incompatible chemicals (4) Exothermic decomposition of thermally unstable or explosive material such as peroxides, e.g. due to temperature rise, overconcentration or deposition on hot surfaces
		Due to an internal explosion arising from formation and ignition of flammable mixtures, mists or dusts	Ingress of air, e.g. due to inadequate purging of equipment at plant start-up, due to loss of nitrogen purge on flare headers, storage tanks, centrifuge systems and driers.	
			Loss of critical inert diluent, e.g. loss of nitrogen padding	
			Failure of explosion suppressants	

Mode of loss	Potential causes		
			Flammable excursion in oxidation processes, e.g. in heat exchangers or long pipe runs
		Due to physically or mechanically induced forces or stresses	Expansion on change of state, e.g. freezing of water in pipe runs
			Thermal expansion of blocked-in liquids, e.g. in heat exchangers or long pipe runs
			Ingress of extraneous phases, e.g. gas compressor failure due to liquid carry-through to machine suction, condensate hammer in steam lines, etc.
	Underpressuring of equipment (for equipment not capable of withstanding vacuum)	By direct connection to an ejector set or equipment normally running under vacuum	Due to equipment malfunction, e.g. loss of liquid seal due to failure of a level controller causing vacuum to be applied upstream
			Due to operator error, e.g. isolation valve left open
		Due to the movement or transfer of liquids	Pumping out of tanks or vessels
			Emptying or draining elevated blocked-in equipment under gravity
		Due to cooling of gases or vapours	Condensation of condensable vapours, e.g. vessel blocked in after steaming
			Cooling of non-condensable gases or vapours, e.g. storage tank by heavy rainfall in summer
		Due to solubility effects, e.g. dissolution of gases in liquids	

Mode of loss	Potential causes	
	High metal temperature (causing loss of strength)	Fire under equipment, e.g. due to spillage, pump leak.
		Flame impingement causing local overheating, e.g. on furnaces due to misalignment or mal-adjustment of burners
		Overheating of electric heaters, e.g. due to failure of high temperature cut-out
		Inadequate flow of fluid via heated equipment, e.g. furnace tube failure on loss of hot oil flow
		Higher flow rate or higher temperature of the hotter stream, or lower flow rate or higher temperature of the colder stream, via a heat exchanger
	Low metal temperature (causing cold embrittlement and overstressing)	Overcooling by refrigeration units, e.g. due to control fault, wrong refrigerant.
		Incomplete vapourization and/or inadequate heating of refrigerated material before transfer into equipment of inadequate temperature rating
		Loss of system pressure on units handling liquids of low boiling point
	Wrong process materials or abnormal impurities (causing accelerated corrosion, chemical attack of seals or gaskets, stress corrosion cracking, embrittlement, etc.)	Variations in stream compositions outside design limits
		Abnormal impurities introduced with raw materials or wrong raw materials
		By-products of abnormal chemical reactions
		Oxygen, chlorides or other impurities remaining in equipment at start-up due to inadequate evacuation or decontamination
		Impurities entering process from atmosphere, service connections, tube leaks, etc. during operation

Appendix 11 Process flow failure modes—"The List"

Equipment	Process failure modes
Entering scope of review drawing	• Pressure increases in incoming stream • Pressure decreases in incoming stream • Temperature increases in incoming stream • Temperature decreases in incoming stream • Incoming stream is contaminated (light ends, heavy ends, salts, chemical additives, pH, etc.) • Incoming stream contains unexpected phases (solids, liquid hydrocarbon or aqueous, vapour)
Control valves	• Fail open • Fail open with bypass open • Fail closed—or partially closed (bypass closed) Note: partially closed valves can present a hazard if there is not enough flow to sustain an effective purge, sweep or to keep a flame lit (flare, furnace, boiler, etc.).
Manual block valves	• Normally closed block valve left open, or opened during normal operation • Normally open block valve left closed, or closed during normal operation (or partially closed) Note: partially closed valves can present a hazard if there is not enough flow to sustain an effective purge, sweep, or to keep a flame lit (flare, furnace, boiler, etc.).

(*Continued*)

Equipment	Process failure modes
Vent, drain and bleed valves	• Vent, drain or bleed valve opened during normal operation • Vent, drain or bleed valve left open at start-up
Emergency (or remotely operated) isolation, depressurising, venting, purging valves	• Emergency isolation valve fails to close when required • Emergency isolation valve fails closed during normal operation • Emergency depressurising, venting or purge valve fails to open when required • Emergency depressurising, venting or purge valve fails open during normal operation
Heat Exchangers (shell and tube)	• Shell side blocked in while tube side flowing • Tube side blocked in while shell side flowing • Tube rupture occurs • Tube leak occurs • Shell side blocked in while exchanger is shut down • Tube side blocked in while exchanger is shut down • Inadequate heat exchange • Excessive heat exchange
Heat exchangers (aerial coolers):	• Excessive cooling occurs in exchanger • Tube rupture occurs • Tube leak occurs • Fan stops due to mechanical or electrical failure • Inadequate cooling due to fouling • Exchanger blocked in during shutdown
Heat exchangers (plate and frame, spiral, etc.)	• Hot side blocked in while cold side flowing • Cold side blocked in while hot side flowing

(*Continued*)

Equipment	Process failure modes
	• Rupture occurs between hot and cold sides • Leak occurs between hot and cold sides • Hot side blocked in while exchanger is shut down • Cold side blocked in while exchanger is shut down • Inadequate heat exchange • Excessive heat exchange
Piping segments	• Piping segment left blocked in with heat tracing on (or off) • Piping segment left blocked in and ambient temperature changes • Any dead legs in this section/node? • Check valve sticks open and forward flow stops • Atmospheric vent line becomes plugged from atmospheric sump, vessel, drum, etc. • Hose rupture occurs or hose becomes disconnected • Expansion joint failure occurs • Restriction orifice plugs • Restriction orifice erodes/corrodes away
Pipelines	• Pipeline leak or rupture occurs
Pumps	• Online pump stops due to mechanical or electrical failure • Check valve sticks open on pump discharge and pump stops • Pump started with suction block valve closed • Pump started with discharge block valve closed • Check valve sticks open on discharge of standby pump with suction and discharge block valves both left open

(Continued)

Equipment	Process failure modes
	• Check valve sticks open on discharge of standby pump with suction block valve closed and discharge block valve left open • Variable frequency drive fails and speeds up (or slows down) the pump • More pumps in parallel service operating than required • Pump seal (packing, etc.) failure occurs • Suction (or discharge) vibration dampener fails on positive displacement pump
Compressors	• Online compressor stops due to mechanical or electrical failure • Check valve sticks open on compressor discharge and compressor stops • Compressor started with suction block valve closed • Compressor started with discharge block valve closed • Check valve sticks open on discharge of standby compressor with suction and discharge block valves both left open • Check valve sticks open on discharge of standby compressor with suction block valve closed and discharge block valve left open • Variable frequency drive fails and speeds up (or slows down) the compressor • More compressors in parallel service operating than required • Compressor seal (packing, etc.) failure occurs • Suction (or discharge) vibration dampener fails on positive displacement compressor

(*Continued*)

Equipment	Process failure modes
Vessels/tanks	• Rate of inflow to vessel exceeds rate of outflow (consider for each liquid phase) • Rate of outflow from vessel exceeds rate of inflow (consider for each liquid phase) • Failure of individual internals (depends on nature of internals—i.e. demister mat plugs, internals collapse and block outlet nozzle, weirs collapse, etc.) • Failure of heating coils or cooling coils • Failure of mixers/agitators • Packing failure of mixers/agitators • Solids accumulate in vessel • Material in vessel ages/decomposes or otherwise changes composition over time (shelf life of chemical in storage, biological growth in diesel fuel, stratification of chemical mixture in storage, other)—generally relevant only for chemical injection tote tanks or storage tanks, but ask the question if unsure
Pressure relief devices	• Pressure relief device sticks closed in dirty/sticky service • Pressure relief device freezes closed (if credible) • Pressure relief device opens and fails to reseat (when discharge is to another part of the process, i.e. pump suction, process vessel, etc., and not easily detected) • Pressure/vacuum device fails to close after operation restored to normal pressure • Rupture disk fails to rupture when required

(*Continued*)

Equipment	Process failure modes
	• Rupture disk ruptures during normal operation (or during upset)
Distillation columns	• Tray collapse occurs • Trays become fouled
Reactors	• Catalyst deactivated/fouled or reaction stops • Catalyst bed plugs off • Excessive reaction rate or runaway reaction • Internals failure • Catalyst residue left in piping during catalyst change-out (watch out for off-gassing while vessel is open, presenting personnel hazards due to toxic gases)
Dryers, molecular sieve units, etc.	• Media becomes plugged • Media is deactivated • Switching valve failure occurs (consider all modes of operation, consider individual valve failures) • Media is too active (or absorbs/adsorbs unwanted components) • Media must be changed or partially removed (bed exposed to atmosphere/vessel entry concerns) • Media residue left in piping during change-out (watch out for gassing off while vessel is open, presenting personnel hazards due to toxic gases)
Fired heaters	• Tube leak or tube rupture occurs • Combustion air supply filter plugs off • Induced draught or forced draught fan stops • Natural draught, induced draught or forced draught damper wide open

(*Continued*)

Equipment	Process failure modes
	• Natural draught, induced draught or forced draught damper closed • Flame arrestor plugs off or is otherwise damaged • Tubes (or other heat transfer surface) fouled • Excessive heat transferred • Insufficient heat transferred • If a bath is used, bath leaks and liquid level lost • If a heating bath is used, medium deteriorates or is contaminated • For fuel supply issues, see valve failure types
Filters/strainers	• Filter/strainer becomes plugged • Media not replaced in filter/strainer after maintenance • Backflow into filter/strainer during cleaning/change-out
Operational	• Normal operating mode • Start-up mode—consider all of the above as required, including cold s/u • Planned shutdown mode—consider all of the above as required • Emergency shutdown mode—consider all of the above as required • Unusual operating modes—vessels bypassed, equipment out for maintenance, etc. • Equipment spacing • Sampling • Ease of accomplishing required tasks (human factors)
Trucks at loading/offloading stations	• Hose rupture occurs or coupling becomes disconnected during loading/offloading

(*Continued*)

Equipment	Process failure modes
	• Truck moves during loading/offloading • Static charge accumulates during loading/offloading • Vehicle collides with truck during loading/offloading • Loading not stopped when truck full (or suction vessel empty) • Offloading not stopped when truck empty (or destination vessel full) • Incorrect or contaminated material offloaded • Material loaded to contaminated or wrong truck • Loading/offloading rate too high • Loading/offloading rate too low
Maintenance	• Equipment with lower-than-desired reliability (especially any identified as safeguarding devices)
Dust or other fine solids	• Static electrical build-up occurs in area where dust is present • Heat accumulation occurs in area where dust is present (friction, other) • Dust explosion hazards in area
Streams leaving scope of the review drawing	• Flow is blocked downstream • Backflow into leaving stream occurs from downstream equipment
System failures (considered only for last stream associated with a given vessel, pump, etc., or group thereof—must be considered for every piece of equipment in the review)	• Pool fire—consider equipment spacing as well as overpressure, etc. concerns • Jet fire—consider equipment spacing as well as fireproofing aspects • Power failure (sometimes local power failure and total failure

(Continued)

Equipment	Process failure modes
	present different consequences, so the team may list more than one unmitigated consequence for this cause, or may want to consider more than one cause)
	• Instrument air failure (sometimes local instrument air failure and total failure present different consequences, so the team may list more than one unmitigated consequence for this cause, or may want to consider more than one cause)
	• Steam failure
	• Cooling medium failure
	• Heating medium failure
	• Heat tracing failure
	• Other utility failure, as applicable (nitrogen, refrigeration, utility air, etc.)
	• Blocked in liquid thermal expansion concerns (in cased any were missed when asked per the piping segment or heat exchanger question sets above)
	• Dead leg concerns in the area
	• Low temperature brittle fracture concerns (carbon steel becomes brittle at temperatures $< -29°C$, which can occur in cold climates or if autorefrigeration is a factor for the process during normal or process upset conditions)
	• Start-up/shutdown issues in the area
	• Equipment spacing/location concerns in the area
	• Emergency isolation (and depressurisation) concerns
	• Maintenance isolation concerns in the area

(*Continued*)

Equipment	Process failure modes
	• Failure of control signal from remote DCS/PLC (if and as appropriate) • Commissioning issues (new equipment)

From Assess Hazards With Process Flow Failure Modes Analysis (McGregor. R.J, Chemical Engineering Progress, March 2013).

Appendix 12 Consequence pathways

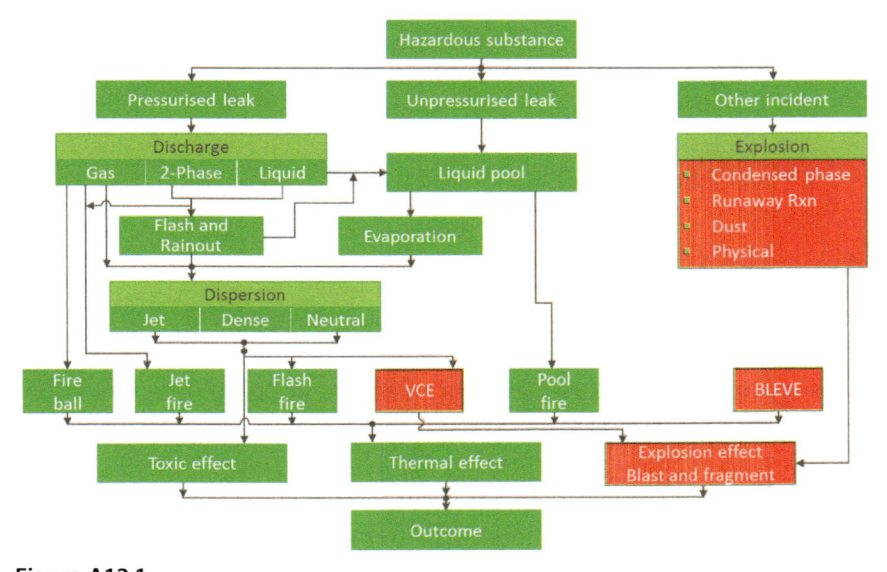

Figure A12.1 .

Appendix 13 Ground rules

Note: Ground rules should be discussed and agreed by the HAZOP team, rather than imposed by the facilitator. The facilitator can present ideas as the basis for discussion and agreement, however.

Possible ground rules relating to methodology:

- We will identify causes in the node but consequences anywhere
- We will describe ultimate unmitigated consequences—what will happen, where and who or what is hurt or damaged
- We will assess consequence UNMITIGATED and frequency including SAFEGUARDS

- We will record a discussion if it lasts more than 5 minutes, even if it does not identify a significant hazard
- We will create an action or parking lot entry if a specific discussion lasts more than 10 minutes
- We will help those who will eventually receive actions by following the rules STANDALONE and WHAT, WHERE, WHY and by quoting tag numbers

Possible ground rules relating to meeting conduct:

- Everyone will contribute
- The scribe is a full member of the team
- We will discuss as a team and not split into smaller meetings
- We will speak one at a time and address the whole meeting
- We will respect everybody's opinion
- We will reach consensus wherever possible
- We will only take calls and respond to e-mails if absolutely necessary
- We will follow up parking lot issues as fast as possible

Appendix 14 Group dynamics

One key role of the facilitator is to identify and influence the dynamics that develop when people interact with one another in HAZOP teams. This appendix summarises a number of aspects of group dynamics in relation to HAZOP [1].

Leadership

Group leadership involves more than your formal position as leader. You cannot rely on your formal authority to see you through the study, because that can be challenged at any time. You have to earn your authority through the respect of team members; they have to feel comfortable with your style of leadership and believe that it is effective.

Other team members may be psychosocially driven to compete with you for the role of "effective leader" (this is discussed further below). Even if they are not, if your style is not accepted by the team then you are likely to be criticised, challenged, ignored or countered and alternative leaders may vie for position, or it may simply result in a leadership vacuum, with the team stumbling onwards without clear direction and consistency.

In a smooth-functioning HAZOP team, team members' enthusiasm can find them taking on aspects of the leaders' role, e.g. guiding the team in the development of a scenario or suggesting additional notes to be recorded. This is not an issue (in fact it is often a feature in very mature teams), but the HAZOP leader should continue to exert their presence at the start and end of each session and during it by means of regular summarising.

Participation

You want people to feel comfortable and willing to fully participate in the meetings as and when required, and you need to monitor that this is happening. Equal participation is often not the norm, and in HAZOP it can be heavily distorted by the team's reliance on the process engineer for their technical knowledge of the process. In spite of this, it is appropriate that all team members are engaged and contributing regularly; as a facilitator you can readily bring team members into the discussion by simply asking for their view on the matter under discussion.

Communication patterns

Understanding the communication patterns within the team can help you understand the underlying dynamics. Communications can be indications of attempts to lead, attempts to dominate or forms of "counter" behaviour. Who talks to whom and who follows who with some verbal or nonverbal expression often indicate alliances, bonds, or influence, or a clear demonstration of differences. The intensity, tone, volume and inflection of speech can indicate disrespect for individuals or the HAZOP process; likewise annoyance or frustration with other individuals or the process as a whole. Nonverbal communication can be positive in indicating agreement or approval (e.g. nodding) but can also betray boredom (sitting back, arms crossed or withdrawal of eye contact), disagreement or surprise (a sudden move forward).

One of the commonest issues in HAZOP is two individuals conducting a side conversation and distracting the rest of the team. This can sometimes be two individuals that are so interested in what is being discussed that they want to continue it rather than move forward (often two technically-minded members), but it can sometimes be on other topics, thereby representing disengagement from the HAZOP activity. In either case it is readily stopped by standing up and moving around the table behind the individuals and then addressed one to one at the next break.

Ideally, every individual communication will be addressed clearly and in a measured way to the whole team; you can facilitate this with the right room setup.

Task and maintenance behaviours [2]

Task behaviours and **maintenance behaviours** are both needed and should be appropriately balanced if the team is to develop into a strong HAZOP team.

Task behaviours help the group identify, develop and analyse hazardous scenarios. They include initiating, proposing or suggesting; building on or elaborating; coordinating or integrating; seeking information or opinions; giving information or opinions; clarifying; recording or capturing content; questioning; disagreeing or challenging; testing for understanding; orienting the group to its task; testing for consensus; summarising.

The HAZOP leader should aim to model or demonstrate these behaviours to the team so that if individuals do not already exhibit them, then they can emulate the example of the leader or other team members. For example, initiating and proposing/suggesting are important when the team is looking for causes of deviations, whereas seeking information, questioning, challenging, building on and testing for understanding are all important in developing and seeking consensus on the descriptions and analysis of hazardous scenarios.

Maintenance behaviours help to develop a team identity and develop social relationships between team members. They include encouraging, expressing feelings, compromising, facilitating communication, interpreting and listening. Such behaviours must be present among the members or else the team will fail to develop into a cohesive, effective unit. The facilitator again models, supports and assists in building these behaviours.

It is an important role of the facilitator to deliberately monitor the group for both types of behaviour, identifying any imbalances or gaps and reacting accordingly.

Psychosocial considerations [1]

Individuals have a variety of needs and styles as well as interpersonal and social preferences which can give rise to a wide range of issues and related dynamics. These dynamics can present in single individuals, between members of the team or across a whole team. Because they are deeply held, they are very difficult to influence for a facilitator, but being able to identify them can help to influence your style of facilitation.

Trust, particularly being valued and respected by others, and **inclusion**, especially feeling recognised or important, are considerations that can arise in HAZOP where teams can be made up of members with wide ranges of social background, ethnicity, rank, qualifications and experience. Leaders need to keep an eye out for individuals who may be uncomfortable in relation to their status in the team and seek to help them to gain the respect of other team members and make a full contribution; this is often required with operating or maintenance technicians or younger team members. Individuals may either seek **control** or **power** which may result in one team member seeking to dominate the meetings or several team members competing for control and challenging your authority. Other individuals may seek to display their **autonomy**, manifesting itself in being unwilling to be influenced, accept the views or opinions of others or form part of a consensus decision.

Individuals can display a **dependency** on other team members, seeking their approval of any contribution they make or **counter dependency**, the need to be seen to be going against the prevailing view of the group, which will make consensus difficult to achieve.

HAZOP is not a precise technique; teams have to progress relatively quickly, making judgments in relation to scenario consequences, e.g. with little or no technical data and dependent on individuals' experience. This can create discomfort for those who have a low **tolerance for ambiguity** or discomfort with uncertainty and HAZOP leaders need to work closely with such individuals to provide them with reassurance that the decisions that are being made are sufficient for the purposes of the HAZOP technique.

Competition, driven by insecurity or the need to prove worth or simply the desire to be seen as better or smarter than others, can manifest itself in individuals taking up valuable meeting time to expound their theories, or apply their experience to an excessive extent in the development of scenarios. Leaders need to be able to acknowledge the worth of individuals while discouraging them from trying to show off; other team members are invariably unimpressed by this type of behaviour.

Finally, HAZOP sessions can be intense affairs with teams working in relatively confined environments for long periods of time. Teams may develop levels of expression, openness or camaraderie which come across to others as excessively **intimate**. Conversely, to others, the atmosphere may be too sterile and insufficiently intimate. Leaders need to be able to spot these tendencies and aim to create a balance that works for the whole team.

Psychosocial-driven behaviours require the most experience and skill to identify and influence since there are often deep underlying patterns causing individuals to behave as they do, to have the needs they have and to deal with others in the way they do. As facilitator your primary interest is in the team working well and progressing; you need to understand these individual and interpersonal issues because they can adversely affect the effectiveness of the study and the enjoyment and happiness of team members.

Decision-making

As facilitator, you can make the determination—on you own or with the team—as to which decision methods will be considered. You can offer to the team the various means of making decisions and help members clarify which method they will use. This decision-method choice can be made before each decision but is more appropriately handled as part of the opening review at the start of the study. In HAZOP, the recognised best practice is consensus decision-making, with the leader exerting authority or making final decisions only in cases of debate over the application of the methodology, e.g. in questions relating to possible double jeopardy situations or "cause outside the node" decisions.

Conflict resolution

Conflict is inevitable, particularly in new teams at the start of the HAZOP study. Heterogeneous groups with a variety of diverse viewpoints, backgrounds, functional interests and expertise will naturally have differences. How such differences are raised, discussed, managed and resolved are critical to the work of the team and in creating trust and an open, safe environment.

The expertise and opinions of the participants need to be shared and used for comprehensive hazardous scenarios to be developed and analysed and effective recommendations for risk reduction constructed. Team members generally learn and produce better results by exploring differences and understanding one another's perspectives. Teams are debilitated and unproductive only when conflicts become fights, positions get cemented, adversaries become hostile toward one another or important points of view are blocked.

It is the facilitator's job not only to safeguard ideas but also to create a safe atmosphere for the open airing of different viewpoints. The facilitator is responsible for helping different parties state their positions, hear one another, engage in balanced, rational dialogue, and involve all team members in resolving issues.

Appendix references

[1] Fast-Track Open Facilitation Skills Workshop Participant Workbook, Facilitate This! 2016.
[2] Hunter D, Training for Change, 2006 http://www.TrainingFor Change.org.

Appendix 15 Facilitation styles and group maturity

The HAZOP leader needs to understand what stage of development the team is at, particularly in the early part of a study. This will help to avoid frustration at the beginning of the study ("are we ever going to get working as a team?") and help the leader to identify when to alter their style of facilitation as the team develops in maturity. The following tables match the stages of development of the team with the role of the facilitator [1] using the descriptors for the stages of development of a group working [2].

FORMING (IMMATURE TEAM)

Team characteristics	Role of the facilitator
Unacquainted individuals	Encourage individual introductions
Uncertainty over the purpose of HAZOP	Invite members to share their concerns and expectations
Uncertainty or nervousness in relation to individual roles within the team ("when should I say something?") leading to low involvement	Clearly explain the objectives of the study
	Clearly explain HAZOP concepts and emphasise as you develop the early scenarios slowly and methodically: check for understanding
Confident or HAZOP-savvy individuals dominate discussions	Identify individuals' styles, strengths and weaknesses
	Rein back dominant individuals by asking others to speak

Maximum support

(Continued)

STORMING (LOW MATURITY TEAM)

Team characteristics	Role of the facilitator
Individuals struggle with key concepts such as cause in the node consequences anywhere, ultimate unmitigated consequences ("why are we doing it this way?")	Encourage individual concerns to be aired and address them with careful explanations, enlisting support from other team members
Conflicts arise, e.g. in relation to credibility of causes and consequences, applicability or adequacy of safeguards ("that could never happen!")	Encourage quieter team members
	Allow conflicts to surface and help to resolve them: build bridges between individuals
	Demonstrate consensus-seeking methods
Frustration with speed of progress (too fast for some; too slow for others)	Frequent pauses to review and cement common understanding
New team member slows development	Maintain calm and patience

High level of support with frequent reviews of progress

NORMING (SEMI-MATURE TEAM)

Team characteristics	Role of the facilitator
Common approach to scenario development is in evidence	Regularly review performance
Scenario identification and development picking up speed	Develop capacity for team to compensate for any individual weaknesses
All team members confident in their contributions	Identify and point out successes and applications of good practice
Minimal conflict	Provide direct support when necessary
Mutual supporting between team members is emerging	
Team comfortable with speed but still may be reliant on facilitator to drive progress	
New team member forces dynamic backwards	

Moderate level of support; identify successes

(Continued)

PERFORMING (MATURE TEAM)

Team characteristics	Role of the facilitator
Full involvement with lots of mutual support and encouragement Identification and development of scenarios initiated and driven by team members Working directly with recorder without leader intervention Efficient progress being made	Allow team members to lead or the group to lead itself Intervene only when absolutely necessary Expose successful team functioning to management and seek their acknowledgement

Minimal support

Appendix references

[1] Fast-Track Open Facilitation Skills Workshop Participant Workbook, Facilitate This! 2016.

[2] Developmental Sequence in Small Groups', Tuckman B.W, Psychological Bulletin, 63, 384–399, reprinted in Group Facilitation: A Research and Applications Journal, Number 3, Spring 2001.

Appendix 16 Communication styles in facilitated sessions

Style	At their best in a session	At their worst in a session	Prevention strategies
Drive	Driving for efficiency Participating Directing Making direct comments Giving the end point first	Alienating by being forceful Not letting people catch up Making snap decisions Killing creativity Being unaware of what's happening with the group	Keep the session fast-paced, well planned Lay out the process and the benefits Get them on your side to go with the flow

(Continued)

Style	At their best in a session	At their worst in a session	Prevention strategies
Influence	Participating Creative Talking Keeping energy up Cheerleading and supporting	Don't stop talking Don't listen Don't want to take time for important details Blue-skying, unrealistic	Give lots of chances to talk Enlist help for out-of-box thinking and getting others to speak Have ground rules: keep discussions relevant, give end point first, avoid bar discussion Take reality check
Steadiness	Friendly Supporting, nodding, agreeing Paying attention listening well Tolerant Acting as peacemakers	Going along with what they don't believe Being the silent martyr Checking out Acting passive-aggressively in response to challenge	Check for agreement Encourage them to challenge but avoid putting on the spot Use their name a lot Reinforce with praise
Compliance	Looking at the details Constructive critiquing Identifying impacts of decision Keeping on task Providing reality checks	Bogging down in details Giving all the reasons why something won't work Not allowing intuitive judgment Holding unrealistic expectations of quality, details	Set the expectation that more detailed analysis will be done outside the session to confirm directional decisions Remind of the level of detail needed for each decision

Adapted from The Secrets of Facilitation by Michael Wilkinson, Jossey-Bass, 2012.

Appendix 17 Managing dysfunctional behaviour

The mobile phone addict

Description	The person's mobile constantly rings, or the person is on and off the mobile frequently or leaves the room frequently.
Common causes	• The person has a high-priority activity that requires attention during the meeting. • The person is unaware of how mobile phone activity can reduce the effectiveness of the meeting for all participants. • The person sees little value in the meeting (probably not a core study member) and is attempting to make the best use of having to be present.
Prevention	• Establish a ground rule; no mobile phone calls during the meeting.
At the time	If a private conversation is possible: • "It looks as though people don't know you're in an important meeting, so they keep interrupting you. Have you been able to get the problem addressed? Is it OK then to turn the phone off or switch it to silent for the rest of the meeting?" If a private conversation is not possible: • "When I heard Phil's phone it was a reminder to me that we need to keep mobile phones off if we can. I want to check with the team to make sure that this won't be a problem."
Later	Discuss the issue privately to ensure that no additional problems exist.

Adapted from The Secrets of Facilitation by Michael Wilkinson, Jossey-Bass, 2012.

The door slammer

Description	The person leaves the room in apparent disgust.
Common causes	• The person has an issue unrelated to the meeting that needs immediate attention. • The person does not believe the meeting is worth investing additional time (probably not a core member). • The person is dissatisfied with the meeting content or meeting progress; perhaps they have felt marginalised within the team or cannot agree with a proposed recommendation.

(Continued)

The door slammer

Prevention	• Establish a ground rule: everyone speaks about issues in the room; we will discuss the undiscussable.
At the time	• Name the behaviour and then spend a few minutes with the team debriefing the event. The debrief helps to form a common group view of the incident: what happened, why it happened, what will be done about it and how we will prevent the rest of us from feeling the need to suddenly walk out. "Wow, Phil just got up and left the room. Given what felt like abruptness, I don't think it was because he had to go to the toilet.""We could try to continue working, but I bet many people are thinking about Phil's departure. So I would like to take a few minutes to get clarity on what just happened. Who can take a shot at explaining what happened and why you think it happened?" "So we have talked about what happened and we have a guess as to why it may have happened. Now I have two other questions. What should we do about Phil? And what needs to happen differently to keep the rest of us from doing what Phil just did?"
Later	• Afterwards, take a break. Meet with the study sponsor or project manager to discuss the issue and select a replacement if necessary. • Follow up to ensure that the agreed-on actions are taken. • Consider meeting privately with the person yourself

Adapted from The Secrets of Facilitation by Michael Wilkinson, Jossey-Bass, 2012.

The quiet one

Description	The person does not participate in discussions.
Common causes	• The person has an introverted communication style and rarely offers comments in group discussions; it could be an operations representative not used to a conference room environment. • The person is typically talkative but is less involved in the discussions because of work pressures or other factors outside the meeting.

(Continued)

The quiet one

	• The person is dissatisfied with what is being discussed or the way the meeting is being run.
Prevention	• Establish a ground rule: everyone speaks.
At the time	• Remind the group of the ground rules (everyone speaks). • Alternatively, employ a brainstorming activity (such as cause identification) to get everyone involved. Begin with someone two or three seats away so that the drop-out will speak third or fourth, to avoid putting the person on the spot and to provide the person with time to prepare an answer. "Let's hear from everyone on this next scenario. I would like to start with [name of person two seats to the right of the drop-out] and go around the table to the left. I'd like each of us to try and imagine how the pressure could rise above 20 barg in this line?"
Later	• Discuss the issue privately to ensure that no additional problems exist.

Adapted from The Secrets of Facilitation by Michael Wilkinson, Jossey-Bass, 2012.

The interrupter

Description	The person interrupts others or finishes their sentences.
Common causes	• The person agrees with the comment made, gets excited and wants to show support. • The person has little patience with the speed with which others speak. • The person feels that what they have to say is more important, or the person disagrees with the comment.
Prevention	• Establish a ground rule: have one conversation; respect the speaker.
At the time	• Validate the interrupter's desire to speak, while transferring the conversation back to the person who was interrupted. "Can you hold that thought for a moment so that Amanda has the opportunity to develop through to the ultimate consequence, then we can add to it? It's hard sometimes when you really want to say something, but let's remember our ground rules."
Later	• Discuss the issue privately to ensure that no additional problems exist.

Adapted from The Secrets of Facilitation by Michael Wilkinson, Jossey-Bass, 2012.

The late arriver or early leaver

Description	The person habitually arrives late to the meetings or leaves early.
Common causes	• The person has meetings or other commitments that make it difficult to arrive on time or stay for the entire meeting. • The person does not believe that the meetings are worth making full attendance a priority.
Prevention	• Distribute the meeting notice in advance; indicate a gathering time of 5–10 minutes prior to the start time; indicate the importance of the purpose and importance for the meeting. • Contact the person in advance to gain commitment to be present for the entire meeting. Get agreement that the meeting should start on time with whoever is present.
At the time	• Remind the group of the ground rules in a respectful way. "I want to thank everyone for being here when you could get here and for continuing to do all you can to arrange your diaries so that we can start on time. Our next deviation is..."
Later	• Discuss the issue privately to ensure that no additional problems exist.

Adapted from The Secrets of Facilitation by Michael Wilkinson, Jossey-Bass, 2012.

The loudmouth

Description	The person dominates the discussion.
Common causes	• The person has an extroverted communication style and is not aware that a tendency to frequently speak first can limit the time and opportunity for others to speak. • The person is aware of the tendency and needs help in balancing time spent talking with time spent listening. • The person intentionally wants to dominate to limit time spent discussing other views.
Prevention	• Establish a ground rule: have one conversation: "share the air." • Meet in advance to let the person know that you will be trying to get others to speak. "I appreciate you being willing to speak, especially given that most have been pretty quiet. I need to get other people speaking more so that we can get their views on the table. So, during this next session, there will be times when you might hear me say, 'Nice point. Let's hear from some others on this'. This way, we'll get everyone's input."

(Continued)

The loudmouth

At the time	• At the start of the next session or scenario, use a round-robin discussion to get everyone involved. Direct the round-robin away from the loudmouth so that everyone else will be able to provide input first. "Let's hear from everyone on this next deviation. I would like to start with [give the name of a person to the left of the loudmouth] and go around the table to the left. 'Can you identify how the flow could be stopped in the feed line to the reactor?'"
Later	• At the break, solicit the person's assistance in getting other people to speak. Remember to empathise with the symptom. Let the person know that from time to time you will purposely not call on them, so that others are encouraged to speak. • Occasionally, make a point of acknowledging the person's desire to speak, but call on someone else. • Follow up to ensure that no additional problems exist.

Adapted from The Secrets of Facilitation by Michael Wilkinson, Jossey-Bass, 2012.

The negative one

Description	The person makes audible sighs of displeasure or negative statements such as "That won't happen," without offering reasoning.
Common causes	• The person has a communication style that focuses on identifying problems. • The person opposes the idea suggested and is identifying reasons for the opposition. • The person opposes the idea suggested and is attempting to create stumbling blocks to prevent adoption. Perhaps they are defensive of their design, facility or team or fearful of potential cost implications.
Prevention	• Establish ground rules: benefits first (i.e. give the strengths of an idea before identifying problems); take a stand (i.e. rather than describe what won't happen, describe what will).
At the time	• Naysayers often express negatively without offering alternatives. Avoid a debate about whether something is wrong by focusing their attention on creating something better.

(Continued)

The negative one

	• Say with optimism, "You might be right. What do you think might happen?"
Later	• Seek to gain agreement to always state benefits before stating problems.

Adapted from The Secrets of Facilitation by Michael Wilkinson, Jossey-Bass, 2012.

The physical attacker

Description	The person physically attacks someone.
Common causes	• Disagreement during the meeting escalates into physical confrontation. • Tensions or issues with a source outside the meeting escalate during the meeting into physical confrontation.
Prevention	• Identify probable issues prior to the meeting: are there personality clashes? • Establish ground rules: discuss the undiscussable; be soft on people but hard on ideas. • Actively keep the conversation focused on seeking solutions rather than assigning blame.

At the time	When a physical attack occurs, it is important to take control.
	  • Stop the meeting immediately.
	  • Let the group know that they will be notified when the next meeting is scheduled.
	  • It is inappropriate to reschedule the meeting then, as a physical attack can restart while the attempt to reschedule is going on.

Later	• Consider meeting with the parties separately to identify the issues and an appropriate course of action.

Adapted from The Secrets of Facilitation by Michael Wilkinson, Jossey-Bass, 2012.

The storyteller

Description	The person likes to tell long-winded stories.
Common causes	• The person has an extroverted communication style and is not aware of the tendency to be verbose. • The person is aware of the tendency and needs help in getting to the point. • The person is aware of the tendency and believes that each story is worth the group's time and should be completely communicated.

(Continued)

The storyteller

Prevention	• Establish a ground rule: "share the air." • Meet in advance to let the person know that you will have limited discussion time in the meeting. "I can readily see how stories of your experiences give people a stronger picture of the point you are making. One of the concerns I have is that I've noticed that sometime people drop out when you begin a story. Is there a way that you can make your end point first and then shorten the story so that most will be able to follow? This may also mean that we can get more done during the meeting. So, during this next session, if I perceive that you may be starting a story, you might hear me say, 'Let's give the end point first so that people will be able to follow you better'."
At the time	Let the person know that he or she needs to shorten the comment. • Stand next to the person and privately, if possible, give the "circling finger" to indicate that the person should wrap up the comment quickly. • Remind the group of the ground rule (state the end point first). "Let's remember the ground rule to give the end point first and keep it brief so that people will be able to follow along better."
Later	• Follow up to ensure that no additional problems exist.

Adapted from The Secrets of Facilitation by Michael Wilkinson, Jossey-Bass, 2012.

The topic jumper

Description	The person frequently takes the group off topic.
Common causes	• The person has a communication style that frequently shifts to a new topic before the prior one is complete.
Prevention	• Establish a ground rule: have one conversation; one topic at a time.
At the time	• Validate the person's point by offering to put it on the issues list or parking lot and then bring the conversation back on topic.

(Continued)

The topic jumper

	"That's a good point. If it's OK, can we put that on the issues list (parking lot) to be discussed in the next node and get back to talking about where we are in the process?"
Later	• Consider seeking an agreement with the person to make an effort to use the issues list (parking lot) when new topics come up.

Adapted from The Secrets of Facilitation by Michael Wilkinson, Jossey-Bass, 2012.

The verbal attacker

Description	The person makes a negative comment about or directed at someone.
Common causes	• Disagreement during the meeting escalates into the verbal attack. • Tensions or issues with a source outside the meeting escalate into a verbal attack during the meeting.
Prevention	• Identify probable issues prior to the meeting; are there personality differences or different views in relation to the quality of the design? • Establish ground rules: discuss the undiscussable; be soft on people but hard on ideas; we are not here to criticise the design but to improve it. • Actively keep the conversation focused on seeking solutions rather than assigning blame.
At the time	• Move between the people to cut off the debate, then slow down the discussion and reestablish order. "Let's take a time-out here. We have important issues to discuss, and we have established ground rules to help us do this. One of our ground rules is to be soft on people and hard on ideas. We are unlikely to be successful if our focus is on blame or finger-pointing; it's in all our interests that we try and make the design as robust as possible. I would like to continue the discussion, if we can, but only if we can do so respectfully and with an understanding of the problems and a focus on developing solutions. Can we do this?"

(Continued)

The verbal attacker

Later	• Consider taking a break and reconvening the meeting later. • Consider meeting with the parties separately to identify the issues and an appropriate course of action. • Consider asking that the person be excluded from future sessions if you believe that the behaviour is likely to persist.

Adapted from The Secrets of Facilitation by Michael Wilkinson, Jossey-Bass, 2012.

The whisperer

Description	The person holds side conversations during the meeting.
Common causes	• The person did not hear or understand a prior comment and asks someone for clarification. • The person heard the prior comment and comments on it to someone. • The person is having an unrelated discussion.
Prevention	• Establish a ground rule: one conversation.

At the time	Take action in a respectful way to let the person know that the behaviour is disruptive.
	  • Move and stand next to the person. Often your proximity to the whisperer is enough to cause him or her to stop.
	  • Privately, if possible, give the "shhh" sign with one finger to your lips if the whispering continues.
	  • Remind the group of the ground rules (respect the speaker).
	"Let's remember the ground rule that we want to have one conversation in the room so that we are respectful of the speaker and other listeners.""If the discussion relates to what we are talking about here, could you share it with the group?"

Later	• Discuss the issue privately to ensure that no additional problems exist.

Adapted from The Secrets of Facilitation by Michael Wilkinson, Jossey-Bass, 2012.

The workaholic

Description	The person does other work during the meeting.
Common causes	• The person has a high-priority activity that requires attention during the meeting. • The person sees little value in the meeting and is attempting to make the best of having to be present.

(Continued)

The workaholic

Prevention	• Establish a ground rule: work only on the meeting in the meeting.
At the time	If a private conversation is possible: • "It looks as though you have some important work to get done, and this meeting has put you in a crunch. We do need your full attention if we can get it. Is this work something you can do later?" If a private conversation is not possible: • "I know we established the ground rule of only doing meeting work during the meeting. I want to make sure that the ground rule will still work for everyone."
Later	• Discuss the issue privately to ensure that no additional problems exist.

Adapted from The Secrets of Facilitation by Michael Wilkinson, Jossey-Bass, 2012.

Appendix 18 HAZOP progress tracker

This very simple progress tracker estimates how much time is required to complete the study based on the "run rate," so far. The average number of nodes completed to-date is projected forward to the remaining number of nodes to estimate how many more days of study are required. This has been proven effective in larger studies.

HAZOP PROGRESS TRACKER	
DATE	26/09/2020
HAZOP PROGRESS	
Total number of nodes	200
Nodes complete	100
Nodes still to HAZOP	100
% completion of HAZOP	50%
HAZOP days to-date	50
HAZOP nodes achieved per day to-date	2.0
Days required to complete HAZOP	50

Appendix 19 HAZOP progress report

HAZOP PROGRESS REPORT														
Client:							Report Date:							
Project														
Subject:							Project No.							
Report				Raised By										

Metrics		Progress by Node			Progress by P&ID									
		Total	Completed	% Complete	Total		Completed	% Complete						
To date:														

Schedule	Original Completion date:	
	Approved Completion date:	
	Forecast Completion date:	
	Progress Update	

Recommendations (S) = Safety; (B) = Business; (E) = Environmental												
Type	S	B	E	S	B	E	S	B	E	S	B	E	Total
Number to date:													

High Risk Recommendations (this period)

Rec No.	Description

Parking Lot Status

Total		Open		Closed	

Parking Lot Summary

Action	Description

Issues and Concerns

Appendix 20 Quality assurance checklist

HAZOP QUALITY ASSURANCE CHECKLIST				
HAZOP Report	�			
Date of Study	▭			
Description of Study	▭			
Reviewer	Role	Signature		Date
▭	▭	▭		▭

1	Was the HAZOP facilitator accredited to lead HAZOPs in this organisation?			Y ▭	N ▭
	Comments:	▭			
	Action Required:	▭			

2	Did the HAZOP Study Team consist of engineering and operational expertise with the following:				
	Understanding of and experience with the Process/facility design and process intent?			Y ▭	N ▭
	• Understanding of and experience with the equipment, design limits, materials of construction and condition of equipment being reviewed?			Y ▭	N ▭
	• Understanding of and experience with the day-to-day operations?			Y ▭	N ▭
	Comments:	▭			
	Action Required:	▭			

			Y	N	N/A
3	Was the following process safety information available, up-to-date and accurate? (if applicable)				
(a)	Piping and Instrumentation Drawings (P&IDs) including:		Y ▭	N ▭	N/A ▭
	• All equipment and systems under study?		Y ▭	N ▭	N/A ▭
	• Equipment and component detail (e.g. pressure relief valve, pump capacity, tank inventory etc.)?		Y ▭	N ▭	N/A ▭
	• Vendor packages if within the scope of the HAZOP?		Y ▭	N ▭	N/A ▭

(Continued)

HAZOP QUALITY ASSURANCE CHECKLIST				
	• Piping specifications?	Y ☐	N ☐	N/A ☐
	• Materials of construction?	Y ☐	N ☐	N/A ☐
	Comments:			
	Action Required:			
(b)	Process Flow Diagrams (PFDs) including:			
	• Heat and material balances	Y ☐	N ☐	N/A ☐
	• Inventories	Y ☐	N ☐	N/A ☐
	• Safe upper and lower operating limits	Y ☐	N ☐	N/A ☐
	Comments:			
	Action Required:			
(c)	Previous HAZID, what if, HAZOP or LOPA studies?	Y ☐	N ☐	N/A ☐
	Comments:			
	Action Required:			
(d)	Control, alarm and trip information including:			
	• Alarm and trip settings?	Y ☒	N ☐	N/A ☐
	• Control system philosophy and description?	Y ☐	N ☐	N/A ☐
	• Interlock/trip activation and response descriptions?	Y ☐	N ☐	N/A ☐
	• Cause and effect diagrams?	Y ☐	N ☐	N/A ☐
	• Emergency Shutdown (ESD) System functions?	Y ☐	N ☐	N/A ☐
	Comments:			
	Action Required:			
(e)	Pressure relief, flare, vent and depressuring information including:			
	• Relief valve datasheets?	Y ☐	N ☐	N/A ☐
	• Scenarios considered for sizing of the relief devices?	Y ☐	N ☐	N/A ☐

HAZOP QUALITY ASSURANCE CHECKLIST				
	• Flare/disposal systems design and sizing information?	Y ☐	N ☐	N/A ☐
	Comments:			
	Action Required:			
(f)	Changes to design since the last HAZOP?	Y ☐	N ☐	N/A ☐
	Comments:			
	Action Required:			
(g)	Standard Operating Procedures (SOPs)?	Y ☐	N ☐	N/A ☐
	Comments:			
	Action Required:			
(h)	Previous process safety accident/incident/near miss reports?	Y ☐	N ☐	N/A ☐
	Comments:			
	Action Required:			
(i)	Process description and process chemistry (if appropriate)?	Y ☐	N ☐	N/A ☐
	Comments:			
	Action Required:			
(j)	Plot plan/layout drawing(s)?	Y ☐	N ☐	N/A ☐
	Comments:			
	Action Required:			
4	Was the HAZOP documented in full to include the following:			
(a)	Recorded in full including:			
	• Node descriptions?	Y ☐	N ☐	N/A ☐
	• Node design intents?	Y ☐	N ☐	N/A ☐
	• Deviations?	Y ☐	N ☐	N/A ☐
	• Credible causes?	Y ☐	N ☐	N/A ☐
	• Consequences (ultimate unmitigated)?	Y ☐	N ☐	N/A ☐
	• Safeguards?	Y ☐	N ☐	N/A ☐

HAZOP QUALITY ASSURANCE CHECKLIST				
	• Risk rankings?	Y ☐	N ☐	N/A ☐
	• Recommendations?	Y ☐	N ☐	N/A ☐
	Comments:			
	Action Required:			
(b)	Have all the hazard scenarios considered by the team for each deviation been recorded, even those with no justifiable cause or hazardous consequence identified, that is, no recording by exception throughout the worksheets?	Y ☐	N ☐	N/A ☐
	Comments:			
	Action Required:			
(c)	Were P&IDs available with nodes clearly marked?	Y ☐	N ☐	N/A ☐
	Comments:			
	Action Required:			
(d)	Were the HAZOP team members' names, expertise and attendance documented during the session?	Y ☐	N ☐	N/A ☐
	Comments:			
	Action Required:			
(e)	Is there evidence that consequences were identified and taken to their ultimate unmitigated consequences?	Y ☐	N ☐	N/A ☐
	Comments:			
	Action Required:			
(f)	Are safeguards (engineering and administrative controls) recorded in the HAZOP work sheets referencing equipment tags?	Y ☐	N ☐	N/A ☐
	Comments:			
	Action Required:			
(g)	Were the Terms of Reference or a Charter documented and approved prior to the Study?	Y ☐	N ☐	N/A ☐
	Comments:			
	Action Required:			
(h)	Are recommendations recorded according to "what-where-why criteria and can they be interpreted without reference to the HAZOP work sheets?	Y ☐	N ☐	N/A ☐
	Comments:			
	Action Required:			

HAZOP QUALITY ASSURANCE CHECKLIST			
(i)	Is there evidence in the HAZOP work sheets that a recommendation was generated when the team determined one of the following conditions existed:		
	• A risk scenario in which the safeguards are unlikely to prevent or sufficiently mitigate a consequence?	Y ☐	N ☐ N/A ☐
	• A significant operability concern?	Y ☐	N ☐ N/A ☐
	• A shortfall in compliance with a regulation or company standard?	Y ☐	N ☐ N/A ☐
	Comments:		
	Action Required:		
5	Were all of the deviations in the following table addressed:		
	• Flow?	Y ☐	N ☐ N/A ☐
	• Pressure?	Y ☐	N ☐ N/A ☐
	• Temperature?	Y ☐	N ☐ N/A ☐
	• Level?	Y ☐	N ☐ N/A ☐
	• Reaction (as needed)?	Y ☐	N ☐ N/A ☐
	Comments:		
	Action Required:		

Appendix 21 Effectiveness assessment questionnaire

Preparation

1. Was the study initiated by the Project Manager; the person with the ultimate accountability for the implementation of the actions?
2. Was the timing of the study appropriate (not too early or too late)?
3. Was the design fully developed?
4. Were the drawings accurate and complete; were they frozen?
5. Was the hazard study leader provided with a clear scope and objectives (terms of reference) for the study by the Project Manager?
6. Was the hazard study leader sufficiently skilled and experienced?
7. Was the study team balanced and chosen to provide the right combination of skills and experience?
8. Were the project, operations and engineering all represented?
9. Were the study team members trained in the hazard study process?
10. Was the study team provided with sufficient notice of the study meetings?

11. Did the study team prepare adequately for the meetings?
12. Was the recording policy for the study defined clearly?

The study meetings

13. Was there adequate time allowed to complete the study thoroughly?
14. Did any individual session last longer than 3 hours?
15. Where the P&IDs, trip logic diagrams, outline process operating instructions (start-up, shutdown, normal), process description, relief philosophy, available and used?
16. Was the boundary of the study clearly agreed and defined?
17. Was the boundary of the study appropriate to ensure that the full impact of the proposed changes could be identified?—including any impact on connected equipment or plant?
18. Were the interfaces between new design and existing plant studied?
19. Was the design intent and design envelope of each part of the study clearly explained?
20. Were all the guide words used, and if not, were the reasons for their omission agreed and documented?
21. Was the study a creative thought process, or did it become mechanistic and boring? Did all team members attend all the meetings?
22. Did the meeting ever conclude that "this will not happen" because we have two protective layers... double jeopardy?
23. Did all team members contribute frequently to the study process?
24. Were contributions open and constructive?
25. Were the meetings suspended in the event that they became bogged-down or boring?
26. Were the actions relevant and clearly defined?
27. Were the actions allocated to appropriate owners, with time scales for completion and a process for reporting back and closing out?
28. Did the study focus on engineering hardware, or did it combine an assessment of hazards relating to hardware with those caused by procedures or human error?
29. Did the hazard study 3 analyse operability issues as well as safety, health, environment and commercial?
30. To what extent did the study modify the design? (see Questions 2–4)

After the study

31. Was the study fully and accurately reported, and was it filed in the Project File with the drawings used in the study meetings?
32. Where the worksheets approved as being accurate and complete?
33. Did the meeting table the conclusions of the HAZOP team where they decided no issue existed?
34. Where the drawings frozen and any ongoing design changes highlighted reviewed by the HAZOP Team and approved?
35. Was an agreed process put in place for approval, review and sign off of HAZOP actions and design changes?
36. Were the actions answered fully and signed off as "accepted" by the HAZOP Team?
37. Were actions that resulted in a change to the design then subjected to the "management of change" (modification or process change request) procedure?
38. Were the actions fed into a tracking process?

Appendix 22 Final report contents

Location	Content	Complete
Main body	Executive summary	After
	Introduction and scope of stud	Before
	Process description and design intent	Before
	Methodology including deviations used	Before
	HAZOP team members and roles	Before/during
	Key issues, risks and recommendations summary	After
	References	During

(*Continued*)

Location	Content	Complete
Appendices	Terms of reference (charter)	Before
	HAZOP worksheets	During
	Key assumptions	During
	List of recommendations with risk ranking	During
	Team attendance by session	During
	Node-identified P&IDs (master set)	During
	Risk matrix	Before
	Observations and drawing changes	During
	Incidents considered	During
	MOCs or modifications reviewed	During
	Other information referenced in the worksheets or used by the team	During
	Glossary of terms and acronyms	During

List of abbreviations

AFD	approved for design
ALARP	as low as reasonably practicable
API	American Petroleum Institute
BDV	blow-down valve
BLEVE	boiling liquid expanding vapour explosion
C&I	control and instrumentation (function)
CED	cause and effects diagram
CV	control valve
CW	cooling water
DCS	distributed control system
ESD	emergency shut-down
ETTO	efficiency thoroughness trade-off
FI(C)A	flow indicator (controller) and alarm
FT	flow transmitter
F(C)V	flow (control) valve
HAZAN	hazard analysis
HAZID	hazard identification (study)
HAZOP	hazard and operability study
HSE	Health, Safety & Environment (also UK Health & Safety Executive)
HX	heat exchanger
IEC	International Electrotechnical Commission
IEEI	inform excite empower involve
IPE	independent process engineer
ISO	International Standards Organisation
LAH	level alarm high
LAL	level alarm low
LI(C)A	level indicator (controller) and alarm
LT	level transmitter
LOPA	layers of protection analysis
MOC	management of change
MOV	motor-operated valve
MPI	main plant item
MSDS	material safety data sheet
NFPA	(US) National Fire Protection Association
NRV	nonreturn valve
P1, etc.	pump 1, etc.
P&ID	piping and instrumentation diagram
P(C)V	pressure (control) valve
PES	programmable electronic system
PFD	process flow diagram
PHA	process hazards analysis
PI(C)A	pressure indicator (controller) and alarm

PICA	pressure indicator controller and alarm
PLC	programmable logic controller
PPSE	professional process safety engineer
PRV	pressure relief valve
PSH	pressure switch high
PSV	pressure safety valve
PSI	process safety information
PSMS	process safety management system
PT	pressure transmitter
QA	quality assurance
RGP	recognised good practice
RR	risk ranking
SMART	specific measurable achievable realistic timebound
SOP	standard operating procedure
S/U, S/D	start-up, shut-down
SWIFT	structured what if? technique
TA	temperature alarm
T(C)V	temperature (control) valve
TIA	temperature indicator alarm
TIC(A)	temperature indicator controller (and alarm)
TP	tie-in point
ToR	terms of reference
TT	temperature transmitter
V1, etc.	valve 1, etc.
VCE	vapour cloud explosion
WWWS	what where why standalone
XCV	high pressure control valve

Index